Chemistry for Geology & Environmental Scientists

By Matthew Pasek, Ph.D.

Chemistry for Geologists and Environmental Scientists by Matthew Pasek
First edition.
Published by Free Radical Consulting
2703 Pemberton Creek Drive, Seffner FL 33584

© 2017 Matthew Pasek

All rights reserved. No portion of this book may be reproduced in any form without permission from the publisher, except as permitted by U.S. copyright law. For permissions contact:

freeradicalconsulting@gmail.com

All rights reserved. This book or any portion thereof may not be reproduced or used in any manner whatsoever without the express written permission of the publisher except for the use of brief quotations in a book review.

Contact us at: www.FreeRadicalConsulting.com

Contents

Prologue 4

Chapter 1. Introduction to Geochemistry 5

Chapter 2. Concept Review 14

Chapter 3. Properties of water 24

Chapter 4. Acids and Bases 32

Chapter 5. Thermodynamics 38

Chapter 6. Redox Reactions 43

Chapter 7. Eh-pH diagrams 50

Chapter 8. Radioactivity 57

Chapter 9. Stable Isotopes 63

Chapter 10. Water-Rock Interactions 1. Metasomatism 70

Chapter 11. Water-Rock Interactions 2. Weathering 75

Chapter 12. Water-Rock Interactions 3. Karst 80

Chapter 13. Water-rock interactions 4. Diagenesis 86

Chapter 14. Properties of Groundwater 90

Chapter 15. Biogeochemistry 94

Chapter 16. Nutrient Cycles 101

Chapter 17. Organic Geochemistry 108

Chapter 18. Biomineralization 118

Chapter 19. Deep-time Geochemistry 124

Chapter 20. Planetary Science: Water in Space 133

Chapter 21. Liquids on other Planetary Bodies 140

Chapter 22. Water as Rock 148

Chapter 23. Meteorites 159

Chapter 24. Instrumentation 168

Prologue

This book came about to complement a course I teach at the University of South Florida, "Fluid Earth 1". When I first began teaching this course, which was supposed to be a hydrology course, I was given free rein to teach the course as desired, so long as fluids were some part of the curriculum. Being a planetary cosmochemist by training, and an astrobiologist by research focus, the course quickly became a geochemistry course with strong doses of planetary science and astrobiology. However, since geochemistry is typically offered as a senior level undergraduate course or a graduate course, and "Fluid Earth" was aimed primarily at junior level undergraduates, it was difficult to find a textbook that provided students with a level-appropriate background in geochemistry.

For five years I provided notes on the course coupled to a thermodynamic database that allowed the students to get raw data for calculations. However, most students expressed a desire for material beyond the notes provided in class, resulting in this text.

This text is effectively "Gen Chem" for Geologists. It is intended as an introduction to geochemistry, but does not go into significant depth in some areas, such as derivations of the laws of thermodynamics and kinetics. Students should take one year of general chemistry prior to using this book, as the basics of chemistry are important to the chapters presented here.

In designing this book, I did not have a production budget, or access to high-quality images. Most of the images here come from my own work, were drawn by hand, or were drawn as schematics in MS Paint or power point, or are public domain from NASA! Please excuse the style- they were done with love if not skill.

Topics covered here include the basic tools of geochemistry including isotopes, thermodynamics, and rate laws, followed by applications of these subjects to geological questions. Additionally, planetary science and biogeochemistry/astrobiology are covered topically near the end of this book, and draw upon the tools provided in the first few chapters. Most of the topics cover water and liquid chemistry, typically at low temperature. High temperature geochemistry, such as that found in petrology, is only touched tangentially, and is reserved for a higher-level geochemistry course.

Matthew Pasek

Chapter 1. Introduction to Geochemistry

1.1 What is Geochemistry?

Geochemistry is the intersection of chemistry and geology. Geochemistry uses chemistry to answer questions in geology. It is an expansive field, and a typical geochemist will use more of the periodic table than the average chemist in his/her considerations. In contrast to chemists, geochemists will tend to focus on the natural chemistry of environmental systems (or the response of those systems to anthropogenic input), and will tend to ignore artificial or lab-based chemistry. Important aspects in geochemistry include abundances of the elements in chemical systems, isotopic studies, redox and acid base chemistry, age dating of samples, mineral formation, water-rock interactions, among many others. This book will provide some of the fundamentals of geochemistry, providing an overview of multiple topics, though primarily in passing. Most of these topics are careers in and of themselves, and a cursory overview cannot do them justice.

1.2 Branches of Geochemistry

Geochemistry was an early branch of chemistry, which, in the early history of chemistry, dealt with the distribution of the natural elements and the common compounds associated with each element.

Economic Geochemistry is one of the oldest branches of geochemistry. This subject deals with finding deposits of valuable minerals and ores. Geochemical tools can assist in this endeavor as the chemical traits of periodicity and compatibility feature prominently in this subject. It is a profitable field, and tends to be what many students think of when they think of making money in Geology. The leading journal in this field is named, appropriately, *Economic Geology*.

Cosmochemistry is the chemistry of places, usually solid and sometimes fluid, outside of the earth. This includes meteorite research, Martian and Planetary chemistry, and some remote sensing techniques. This subject is not "geo"-chemistry, since it does not deal with the subject of the earth. However, tools and techniques developed here have both influenced and been influenced by terrestrial geochemistry. One of the leading journals in this field, and in several of the others below, is *Geochimica et Cosmochimica Acta*.

Isotopic Geochemistry uses isotopes (atoms with the same number of protons, but differing numbers of neutrons) to make estimates on the origin, processing, temperature, and features of a chemical system. Radioactive dating is a tool in isotopic geochemistry.

Organic Geochemistry explores the natural chemistry of carbon within the environment, focusing primarily on no longer extant life (the natural chemistry of carbon within extant life falls under the purview of biochemistry). Organic chemistry includes petroleum geochemistry, and environmental remediation, as well as less applied science fields in paleontology.

Aqueous Geochemistry is the chemistry of water, and of water-rock interactions. As water is one of the major resources for civilization, accurate understanding of the chemistry and quality of water is a critical part of geochemistry.

Environmental Geochemistry includes studying the fate of contaminants, recycling of nutrients, and proper disposal of pollutants with time. This subject is very useful for regulatory agencies. One of the leading journals in this field is *Environmental Science and Technology*.

Biogeochemistry is the intersection of geochemistry with biology. Biogeochemistry includes studies of how organisms are altered by their geochemical environment, and how they in turn alter their geochemical environment.

1.3 Career Opportunities in Geochemistry

As in the general science of geology, there are three main fields to pursue with coursework in geochemistry: research (academic), government, and industry. The activities of a geochemist will vary based on location and job demand. The educational requirements, freedom of inquiry, and pay also vary (Table 1.1).

Table 1.1 Career opportunities with features by group.

Path	Activities	Benefits	Disadvantages	Education
Industry	Resource identification, consulting for environmental and other concerns, remediation	Has the highest salary potential, can become own boss after a while	Must work to win contracts, not always a lot of leniency in research activities	At least Bachelor's, often Master's, and depending on state, a professional geology certificate is beneficial
Government	Permitting, regulation, resource identification, environmental concerns	Has good security, decent pay, some leeway in research endeavors	Never extremely high income, often limited in research foci, job sometimes changes with politics	Sometimes Bachelor's, usually Master's, occasionally Ph.D.
Academia	Research, teaching, and service in a university setting	Has excellent freedom, often has decent security post-tenure, can have good pay	Never extremely high income, universities have lots of bureaucracy, many demands from different offices. Tenure process can be stressful	Ph.D. usually, Master's OK for community college

1.4 Speciation, Abundance and Time

Geochemistry is concerned with three principles aspects of chemistry (Figure 1.1). <u>Abundance</u> tells you what is within a sample. How much oxygen does it contain? How much arsenic? How much is too much? How much is enough to make a profit? <u>Speciation</u> is the bonding structure of the compounds. What's the difference between quartz and stishovite? Aragonite and calcite? Arsenate and arsenite? Speciation and abundance influence each other, that is, speciation can determine abundance (ferrous iron is more soluble than ferric iron, and hence iron will be more common in reducing environments), and abundance can determine speciation (too much oxygen makes all iron ferric).

In addition to speciation and abundance, <u>time</u> features prominently in geology. How long has a compound existed? How old is a rock? When did a geochemical change occur? These questions make a geochemist a geochemist.

Figure 1.1. Diagram summarizing the differences between abundance, speciation, and time.

1.5 Classification of the Elements

Geochemists have long sought to classify the elements according to their chemical behavior. Unfortunately for the student of geochemistry, there are many ways to do so, and not all are completely obvious.

Terrestrial Geochemical Classification: the terrestrial geochemical classification was originally based on the material an element went into when placed in a smelter. A smelter is an apparatus that, when used, attempts to make a metal from an ore. This process takes an ore (for instance, Fe_2O_3 or hematite) and places it in a hot crucible with sand and charcoal. The charcoal reacts with the ore to remove oxygen, making a metal. Typically, four substances would separate during this process. The metal would consist of several metal-loving elements, and elements that bond to the metal would hence be called *siderophile* elements (Figure 1.2). Siderophile elements include nickel, platinum, and iridium. Most transition metals are siderophile elements. Above the metal would be a lighter (less dense) layer of sulfides. These sulfides are those elements that preferentially bonded to sulfur, and are called *chalcophile* elements. Chalcophile elements include tin, lead, and zinc. Chalcophile elements tend to be those that lie to the right of the transition metals, including the elements near to the metalloids. Within the smelter, the sand would act to remove those elements that like to bond to silicate materials, and these elements would be called *lithophile* elements. Lithophile elements include sodium, aluminum, and calcium. This product is also a waste material called *slag*. Alkali metals, alkali earth metals, and metals to the left of the main transition metals, and the rare earths are all typically lithophile elements. Finally, some elements would be vaporized completely, and would escape the smelter as gas; these are called the *atmophile* elements. Atmophile elements include nitrogen, neon, and helium, and generally include the elements you would typically consider as being gaseous. Notably this division is shown in the bulk chemistry of the earth: siderophile elements are found at the core, surrounded by a mantle of lithophile elements.

Figure 1.2. Overview of the smelting process, giving, from bottom to top, the siderophile metal product, chalcophile and lithophile waste products, with escape of the atmophile elements.

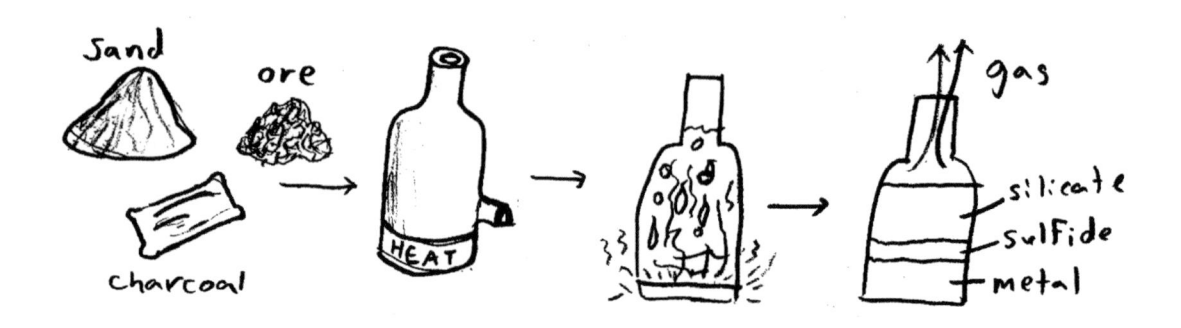

Cosmochemical Classification: space chemistry has its own set of rules for classifying elements. The cosmochemical classification of the elements is based on the temperature at which a given element forms its first solid. At a very high temperature, all elements are gases. The pressure of space in a gaseous nebula is also low (about 0.01% of the atmosphere you are breathing). As this gas cools, elements start to form solids. The elements that form the first solids are called the *refractory* elements (Figure 1.3), and include calcium, aluminum, and titanium. After this form two major groups of elements, the *magnesium silicates*, and the *metals* solidify. Magnesium, silicon, and iron fall within these categories. After these two major, planet-forming materials solidify, some elements will react from the gas to bond with these elements, and these elements are termed the *moderately volatile* elements. The moderately volatile elements include phosphorus, sodium, and potassium. Moderately volatile elements will continue to solidify until the element sulfur solidifies as iron sulfides, formed from reaction of H_2S gas with iron metal. This occurs at about 750 K. Below this temperature, all elements that solidify are called *highly volatile* elements, and include gases such as carbon-bearing gases (CH_4), ices, and the noble gases.

Figure 1.3. Diagram of the cosmochemical classification of the elements.

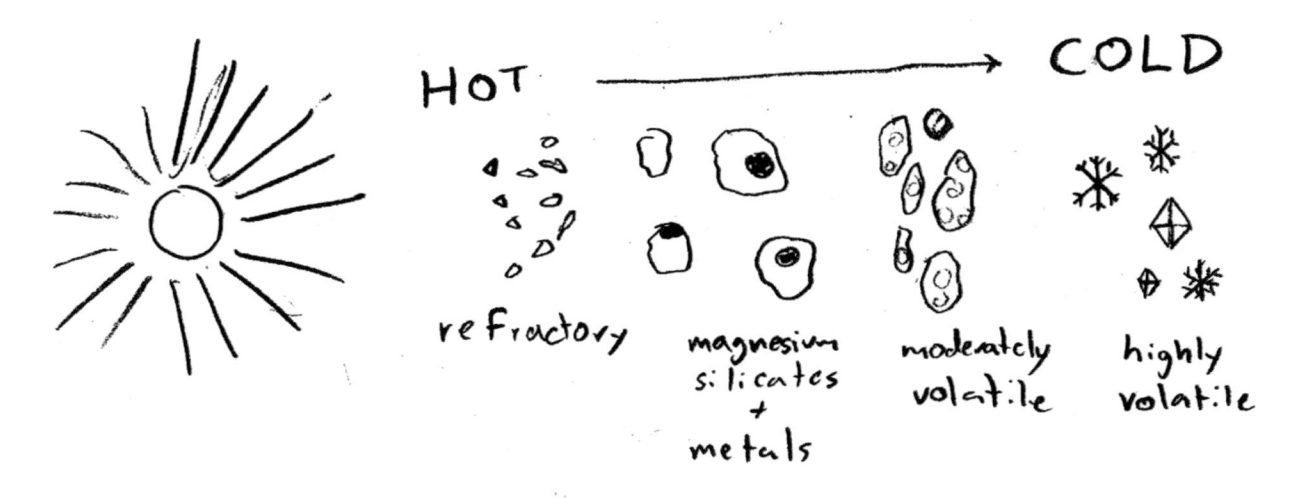

Hydrochemical Classification: a final route to classifying the elements is to place the elements into categories according to their behavior with increasing salinity. If an element becomes more abundant as salinity (Figure 1.4) increases, it is said to be *conserved*. Sodium and chlorine are conserved elements in ocean water- as salinity increases, Na and Cl also increase in abundance. If the element does not become more abundant, it is *non-conserved*. It can be non-conserved for several reasons. If biology is the cause for it not being conserved, then it is not conserved due to biology (*type II*). Type II elements include phosphorus and iron. Otherwise, it is not conserved due to some other reason (*type III*). Lead is a type III element. The reasons why some elements are not conserved may be due to mineral precipitation, reaction with rocks, or photochemistry, among other possibilities.

Figure 1.4. Diagram of the Hydrochemical classification of the elements.

Alien: I come for your gold, human!

Human: But...

Human: You're an advanced, space-faring civilization... Wouldn't it be easier to mine asteroid cores? Gold is a siderophile element, after all.

Alien: Hmm... I hadn't thought of that... I suppose that would be much easier!

Geochemistry For The Win!!!!

Chapter 2. Concept Review

2.1 Review

In this chapter, we will go over key things that you should already be familiar with. These include basics of geology, such as minerals commonly encountered, rules of math (units, sig figs, logarithms), and how to read the periodic table. Hopefully much of this will be familiar or even trivial, if not, review will be necessary.

2.2 Minerals

A mineral is a naturally occurring, crystalline substance of defined chemical composition. Rocks are composed of minerals, but minerals are not composed of rocks. Mineralogy and geochemistry are intimately related; several fields of geochemistry study minerals and the processes by which they were formed. The formation of specific mineral is dependent on many things: composition of the source material, time available for mineral formation, temperature, pressure, and the presence of neighboring minerals.

The most common minerals found on the surface of earth are silicate minerals, in which the SiO_4 tetrahedron features prominently in the crystal structure. In addition, several other mineral groups are quite important to surface geology, including carbonate minerals, oxides, and sometimes sulfides.

Familiarity with minerals and their chemical formulae is a useful for problem solving in geochemistry. Table 2.1 has a list of a few commonly used minerals and their formulae. This is by no means an exhaustive list, and for that, a mineralogy course is necessary.

Reminder

- A mineral is a **naturally occurring**, **ordered solid**, generally consisting of a **defined composition** and generally **inorganic**
    - **So no man-made materials here**
    - It can't be random in structure- no coal
    - **No ultracooled liquids allowed**
    - **Must not vary much between different occurrences**
    - Should not be carbon-based (though exceptions exist to this rule)- so sugar isn't a mineral

Table 2.1 Minerals common in geology.

Mineral Name	Formula
Andalusite	Al_2SiO_5
Apatite	$Ca_5(PO_4)_3(OH,F,Cl)$
Aragonite	$CaCO_3$
Calcite	$CaCO_3$
Corundum	Al_2O_3
Diamond	C
Dolomite	$CaMg(CO_3)_2$
Enstatite (orthopyroxene)	$(Mg,Fe)_2Si_2O_6$
Feldspar	$KAlSi_3O_8$ to $NaAlSi_3O_8$ to $CaAl_2Si_2O_8$
Fluorite	CaF_2
Graphite	C
Gypsum	$CaSO_4 \times 2H_2O$
Halite	NaCl
Hematite	Fe_2O_3
Kyanite	Al_2SiO_5
Magnetite	Fe_3O_4
Olivine	$(Mg,Fe)_2SiO_4$
Pyrite	FeS_2
Quartz	SiO_2
Sillimanite	Al_2SiO_5
Talc	$Mg_3Si_4O_{10}(OH)_2$
Topaz	$Al_2SiO_4(F,OH)_2$
Zircon	$ZrSiO_4$

2.3 Significant Figures

Hopefully you remember sig figs from Gen. Chem. 1. Significant figures are a useful technique to show error in your calculations. When the exact error of a process is unknown or is otherwise unclear, significant figures can be useful for stating the precision of your answer. For example, stating that the mass of a sample is 13.2552 g states that you know the mass to the 0.1 mg place. To be succinct, sig figs are used to express error. A more complex discussion of error in measurement is beyond the scope of this material. Just remember to pay attention to your sig figs!

2.4 Logarithms

Logarithms are exponents. That is, for the equation $A^X = Y$, the $\log_A Y = X$. Geochemistry is a logarithmic science. That is, many chemicals will vary in abundance over several orders of magnitude. One very important logarithm in geochemistry is the pH. The pH is the negative log base ten of the hydrogen ion concentration in water ($-\log_{10}[H^+]$). This number frequently varies over 12 orders of magnitude across the waters and other fluids of the earth. Several rules are associated with logarithms (Table 2.2), and familiarity with these is helpful for solving the algebra behind many geochemical problems.

Table 2.2 Rules of logarithms

$$\log_A Y = X \leftrightarrow A^X = Y$$

$$\log_A(XY) = \log_A X + \log_A Y$$

$$\log_A(X/Y) = \log_A X - \log_A Y$$

$$\log_A(X)^Y = Y \times \log_A X$$

$$\log_A X = \log_B X / \log_B A$$

$$\log 1 = 0$$

$$\log_A A = 1$$

$$A^{\log_A X} = X$$

2.5 Scientific Notation

Scientific notation is a way of writing large and small numbers that allows for easy comparison of magnitude. For instance, comparing 0.000000005 and 0.0000000007 to see which is larger is difficult. However, comparing 5×10^{-9} and 7×10^{-10} is much easier (5×10^{-9} is larger here). Scientific notation can also allow for easy math. For instance, the mass of 1 km^3 of liquid iridium is equal to:

$$M = \rho V$$

Where M is the mass in kg, ρ is the density (20000 kg/m^3), and V is the volume (10^9 m^3). This provides a mass of:

$$M = 2 \times 10^4 \times 10^9 = 2 \times 10^{13} \text{ kg}$$

Note: Scientific notation is a tool that provides the significant figures in the number preceding the 10. That is, 3.16×10^9 has 3 significant figures. It is not permissible to provide this number as 3.16E9, 3.16e^9 or 3.16^9. The latter two are wrong, and the first is sloppy.

Although you don't need to use scientific notation for everything (describing 12 as 1.2×10^1 is just obnoxious), it is useful for any number greater than 1000 or less than 0.001.

2.6 Unit Conversions.

In solving scientific problems, most scientists will use SI units. These are based on the kilogram (kg), meter (m), and second (s) (Table 2.3). You may encounter cgs units, consisting of centimeters (cm), grams (g), and seconds (s), frequently when dealing with engineers. Alternatively, in water science several imperial units are used including feet (ft), pounds (lbs), and gallons (gal).

Table 2.3 SI Units

Unit		Measures
meter	m	Length
kilogram	kg	Mass
second	s	time
	m/s	velocity
Newton	kg m/s^2	force
Joule	kg m^2/s^2	energy
Watt	kg m^2/s^3	power
Pascale	kg /s^2 m	pressure
Kelvin	K	temperature
	kg/ m^3	density
	m/s^2	acceleration

For the most part, units can be transformed from one to the other by simple multiplication. I find google to be a useful unit conversion device:

Example: how many meters are in 44.5 feet?

$$44.5 \text{ ft} \times 0.3016 \text{ m/ft} = 13.4 \text{ m}$$

If water flows through a region at 81 ft^3/hr, what is this in m^3/s?

$$81 \text{ ft}^3/\text{hr} \times (0.3016 \text{ m/ft})^3 \times 1 \text{ hr} / 3600 \text{ s} = 6.2 \times 10^{-4} \text{ m}^3/\text{s}$$

Note here that to transform ft^3 to m^3, you must take the conversion and cube it. Be careful when you do this as there are many times when the order of operations matters for doing this calculation properly!

Temperature is a different conversion, and there are 3 main scales: The Kelvin scale (the SI unit), degrees Celsius (°C), and Fahrenheit (°F). Although only Celsius and Kelvin are used in science outside of the US, the °F is a useful scale for weather as it is based on a one-hundred-point scale applicable to typical weather conditions faced by people (or, more specifically, Europeans).

0°F is quite cold and 100°F is quite hot. Celsius is based around the chemistry of water with 0°C being the freezing point of water, and 100°C being its boiling point. A temperature outside of 40 °C would be quite hot, and -15 °C would be quite cold (and are close equivalents to these qualitative descriptors on the Fahrenheit scale). These scales are not preferred for use in science as they are not absolute scales. For this we adopt the Kelvin scale, which is based on the minimum temperature any object can get, 0 K. Note that Kelvin does not use a degree sign. Many problem-solving techniques in science use Kelvin. Kelvin uses the same relative temperature scale as Celsius, and hence a difference of 50 K between the temperature of two objects is equivalent to a 50 °C temperature difference.

You can convert Fahrenheit to Celsius:

$$(T\ (°F) - 32) \times 5/9 = T\ (°C)$$

Or

$$(T\ (°C) \times 9/5 + 32) = T\ (°F)$$

Add 273.15 to turn °C to Kelvin.

2.7 Chemistry

The periodic table of the elements is a useful tool in geology. An element is typically represented on a periodic table as its chemical symbol (e.g., Ru). This chemical symbol is a shorthand way of describing the element. Familiarity with what the symbols mean (e.g., Au is gold, Pb is lead, Si is silicon) is helpful. The atomic number of element is the number of protons an element has (e.g., 44), and determines its place on the periodic table. Finally, the atomic weight of the element is a weighted average of the isotopes that occur of this element on the surface of the earth (101.07). In other word, this number tells you the mass of one mole of ruthenium.

ruthenium
44
Ru
101.07

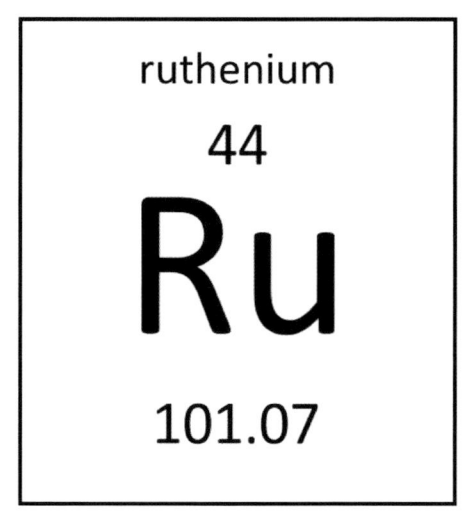

The mole is the final unit we use commonly in geochemistry. The mole is defined as the number of atoms present in 12.0000… grams of isotopically pure carbon-12. A mole is equal to 6.022141×10^{23} atoms. There are more sig figs known to this number than this, and use as many as needed to preserve the error integrity of your calculations (so if you know mass to 4 sig figs, use at least 4 sig figs to this number!). A mole is simply a number, more specifically Avogadro's number. If you would like assign units to it, you can use 6.02×10^{23} particles/mole, and will be accurate to three sig figs. Particles can be atoms, molecules, ions, grains of sand, or even small mammals also called moles, whatever unit you're interested in. Moles are also abbreviated mol (at the loss of one e: this is not a major time-saver!).

One of the major rules of chemistry is that elements seek to obtain a number of electrons that matches a nearby noble gas. The noble gases are helium, neon, argon, krypton, xenon, and radon and are among the least reactive substances on the periodic table. As a result, elements will seek to lose or gain enough electrons to match the electronic configuration of the noble gases (figure 2.1). For instance, silicon will typically lose 4 electrons to become like neon, whereas chlorine will gain one electron to become like argon. The propensity of an element to seek electrons is related to its electronegativity.

The periodic table is divided into rows and columns. Rows of elements share similar densities and abundances, whereas columns of elements share chemistry.

In addition to these features, columns share common charges. For instance, the alkali metals (Li, Na, K, Rb, Cs, and Fr) all possess a common charge of +1. These elements seek to lose one electron to become a positive ion.

Figure 2.1. Elements seek the noble gas configuration. The numbers below each element denote how many electrons they will lose (negative) or gain (positive) to attain the noble gas configuration of electrons.

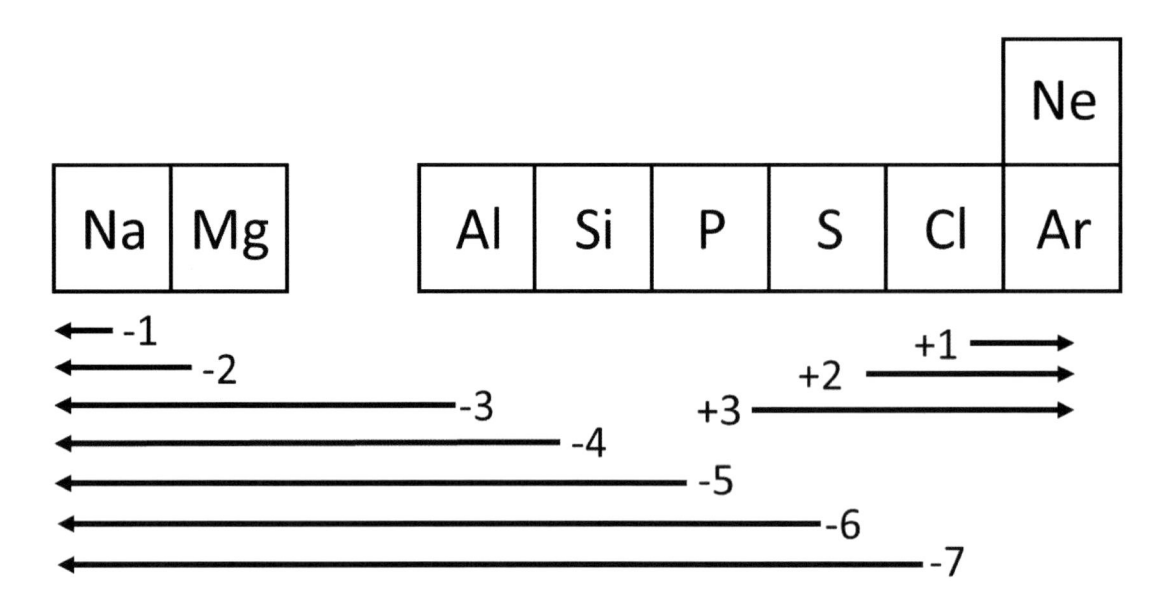

Chapter 2 Practice Problems

1. Convert 64 feet to meters, given 1 foot = 0.3048 m.
2. Convert 60 watts into the equivalent imperial unit (slugs, feet, seconds)
3. If the temperature outside is 68° F, what is this in Celsius? Kelvin?
4. What is $(Q^{90})^3$? $ZRQX^3R^5Z^3/Q^4$?
5. What is $\log_7 49$ (without using a calculator)?
6. Rewrite $-1.7 \log m = \log f$ to get rid of the exponents
7. What do you predict the oxidation state(s) of selenium (Se) will be?
8. What would the mass of a mole of moles be, assuming a mole (mammal) has a mass of 30g?

Chapter 3. Properties of water

3.1 Why Water?

Water is one of the most important molecules on the earth. Its chemistry makes the surface of the earth habitable, drives heat exchange and weather, and is an active part of the rock cycle. The versatility and importance of water to geochemical systems is due to several important characteristics of the water molecule.

Figure 3.1 Diagram of 2nd row hydrides

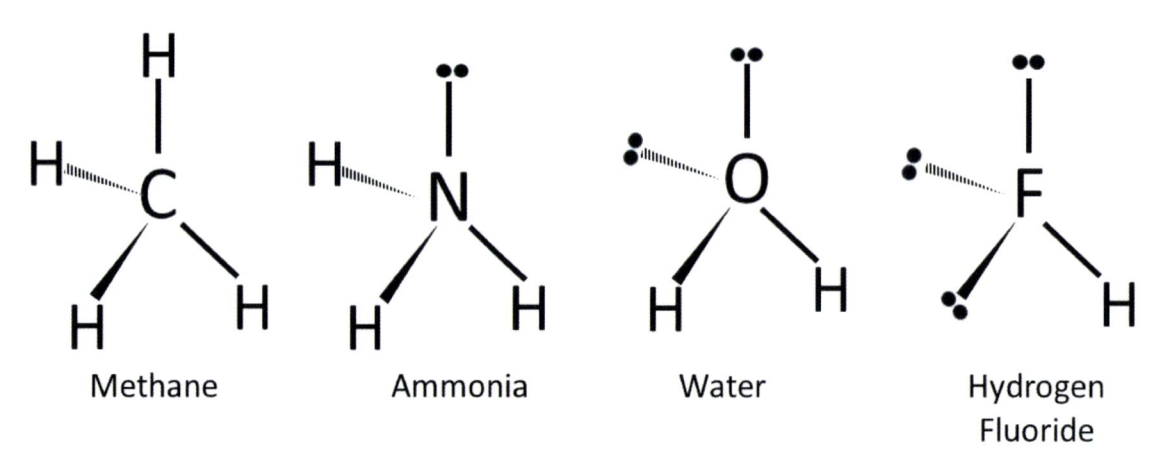

Methane　　　Ammonia　　　Water　　　Hydrogen Fluoride

There are several 2nd row hydrides that exist in nature (Figure 3.1). Many of these are important on a planetary scale (methane, ammonia, and water), depending on location in the solar system. These hydrides vary significantly in physical characteristics (Table 3.1). The reason for these differences is primarily hydrogen bonding. Methane, bearing no free electron pairs, is incapable of hydrogen bonding. Ammonia bears three hydrogens and one free electron pair. Water bears two of both. HF bears three electron pairs and one hydrogen. As a result, water with its 2:2 ratio, has the highest capacity for hydrogen bonds, hence water has the highest temperature for its melting and boiling point transitions.

Table 3.1 Physical properties of hydrides at 1 atm.

Hydride	Melting Point (K)	Boiling Point (K)
Methane	90.7	110
Ammonia	190	240
Water	273	373
HF	190	293

Because of having two positively charged hydrogens and negatively charged free electron pairs on the oxygen, water has a dipole moment. Water is hence a superb solvent for many compounds and rocks.

3.2 Abundance.

Some alternative materials may also be able to dissolve compounds and rocks. However, in this case, it is a question of abundance that shows the dominance of water as a solvent throughout the universe. Alternative solvents that have been considered include formamide, with a molecular formula H_3CNO. However, the abundance of the elements (Figure 3.2) shows that water, consisting of the most abundant element (H) and the third most abundant element (O) is much more likely as a cosmic molecule than formamide that consists of three abundant Hs, and the three less abundant C, N, and O, implying its spontaneous formation is less likely. Indeed, water is the most likely molecule to form in the universe after H_2.

Figure 3.2. Cosmic abundance of the elements from Anders and Grevasse (1989)

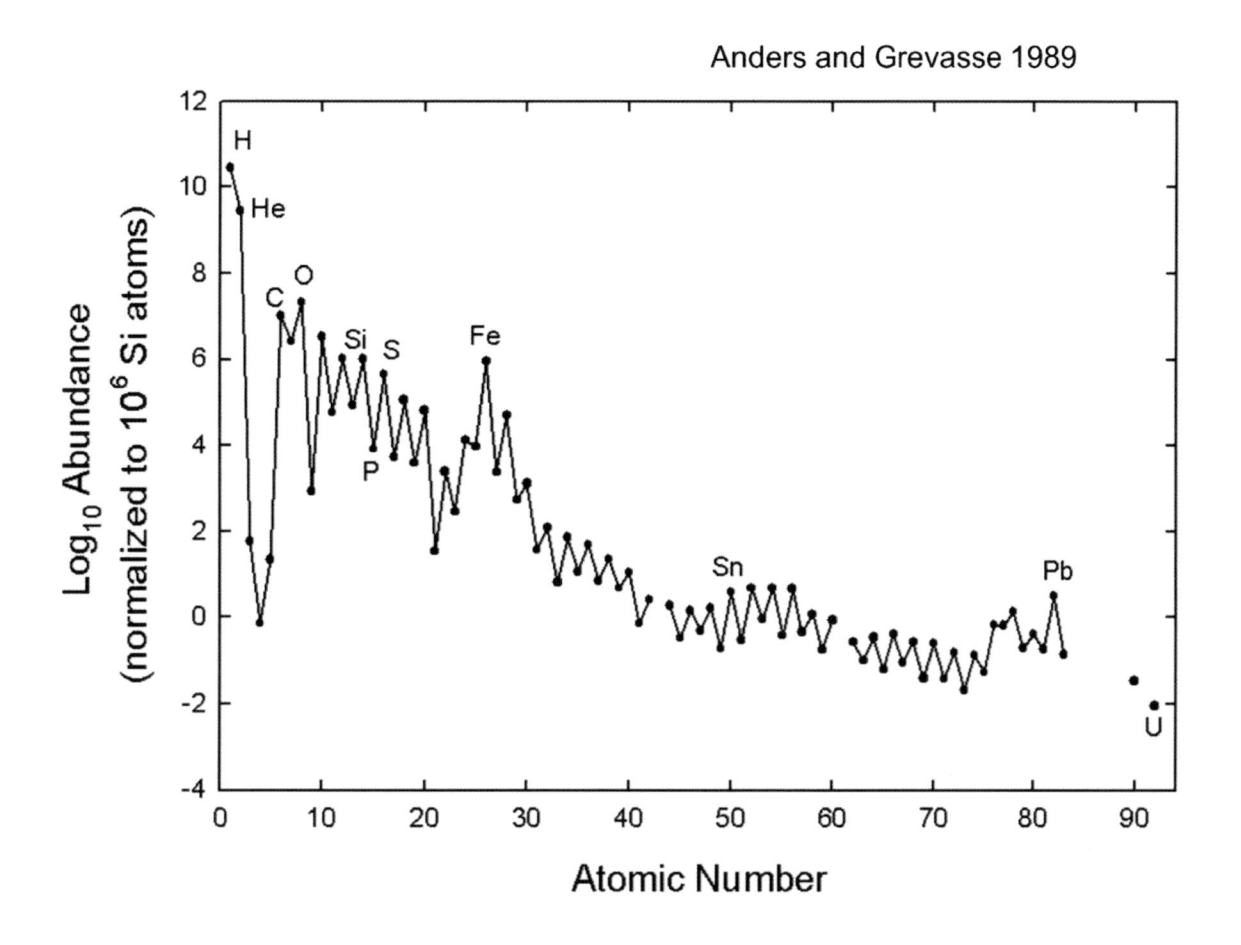

3.3 Thermodynamics.

Water has several thermodynamic properties that allow for the easy calculation of energy required for heating or for phase transitions. One specific unit used exclusively for water is the calorie. The calorie is the unit of energy required to heat one gram of liquid (between temperatures 273-373 K) water by one degree Celsius or Kelvin (these two are identical). This differs from the food "Calorie", which is 1000 calories. A calorie is equal to 4.184 joules.

The heating of liquid water from one temperature to another has an energy requirement equal to:

$$Energy = C_P \times m \times \Delta T \qquad (3.1)$$

where the energy is equal to the heat capacity of water (1 calorie per gram per degree C or K changed), m is the mass of the water being heated or cooled, and ΔT is the temperature change of the water.

The phase changes of water also have specific energy requirements:

$$Energy = m \times \Delta H \qquad (3.2)$$

where ΔH is the heat of fusion or of vaporization of water. For the water-ice transition, the ΔH is equal to 80 cal/g, and for the water-steam transition, the ΔH is equal to 540 cal/g. The transition from a liquid to a gas usually requires more energy than from a solid to a liquid for most molecules.

When water is mixed with salts, the salts dissociate and dissolve within the water to some extent. As a result, the water is now a solution, and has properties that vary slightly from pure water. One of the strongest variations is in freezing and boiling temperature changes. These can be calculated as:

$$\Delta T = k \times m_{solutes} \qquad (3.3)$$

Here the ΔT is the phase temperature change (either decrease of freezing point or increase of boiling point), k is a constant (equal to 1.86 K kg/mol for the freezing point depression, or to 0.512 K kg/mol for the boiling point elevation), and $m_{solutes}$ is the molality of dissolved constituents within the water. If a salt dissolves in water, then the $m_{solutes}$ is the sum of the molalities of each individual dissolved ion in water. This is used extensively to promote the melting of ice in cold climes (i.e., salting the roads).

3.4 Density

The density of water is nominally 1 g per cubic centimeter, or 1 kg per L, or 1000 kg/m^3. These units make water mass calculations easy when given a volume. Water is also an unusual compound in that its solid state is less dense than its liquid state. This results in ice floating on top of water, which has a variety of consequences. Furthermore, the density of water increases as it warms from 0 to 4 °C. As a result, ice melt water is lighter than water that has had a chance to warm. These properties have significant consequences for lakes and other water bodies that freeze during the winter.

3.1 Water Cycle.

The water cycle dominates geology. It consists of the evaporation, respiration, and transpiration of water, condensation of water (in precipitation), surface and groundwater flow of water, and storage of water in the ocean (Figure 3.3). 97% of the earth's water is in the ocean, and of the remaining 3%, about 75% is in ice, with the remainder in groundwater and surface water. The water cycle drives the surface geology and chemistry of the earth, and is the reason why the surface of the earth looks so young compared to the Moon and some other planets.

Figure 3.3. Schematic of the water cycle. The water cycle is characterized by evaporation & transpiration, transportation, and precipitation.

3.2 Dissolution

Water is one of the most effective solvents known. A solvent is a compound that is capable of dissolving another. The compound being dissolved is called a solute, and the resulting mixture is a solution.

The amount of dissolved substances in a water sample is termed the total dissolved solids (TDS). TDS has units of mg/L. TDS is measured through two ways: a sample of water is evaporated to the point where solids are all that remains, and these solids are then massed, or alternatively, the ions are inferred from the electrical conductivity of the water. The TDS is one of the easiest ways of identifying if a water source is potable. In general, a water sample with a lower TDS is safer to drink than a water with a higher TDS.

Table 3.2. Common TDS values for water

Type	TDS
Fresh	<1000
Brackish	>1000 - <10000
Saline	10000
Average Seawater	35000
Brine	>100000
Potable	<300 - 450
Precipitation	10
Surface Water	100
Groundwater	350

The main dissolved solid constituents of a groundwater or surface water are the four cations Mg^{2+}, Ca^{2+}, Na^+, and K^+, and the four anions SO_4^{2-}, Cl^-, HCO_3^-, and CO_3^{2-}. In addition, nitrate (NO_3^-), phosphate, and iron sometimes are important. In addition to these inorganic constituents, most waters also have some amount of dissolved gas, microbial cells, viruses, and organics as materials within the water.

3.3 Units of water

Water chemistry typically uses three sets of units: ppm, mg/L, and M (molarity). These are interconvertible if the TDS of the water is known. The most commonly used unit in groundwater and surface water studies is mg/L, or mg of solute per liter of solution (water) present. In other words, the mass of a solute divided by the volume of water it is dissolved in provides these units. In most cases, this unit is effectively identical to the parts per million of the solute (ppm). Since most waters are dilute, and a liter of water has a mass of one kg, a mg per kg is usually a part per million in weight. Molarity is also used and is equal to the moles of solute per L of solution. This one is most useful in water chemistry as it is directly related to the chemical activity of a solute.

Other units that are used in aqueous chemistry include molality, which is moles of solute per kg of solvent (water), and meq/L, which is the molarity of a solute multiplied by its charge.

Molarity, mg/L, and ppm are interchangeable by the following calculations:

$$\frac{mg}{L} = M \times MW = \frac{mg}{10^6 + TDS} ppm \times 10^6 \qquad (3.4)$$

where the mg of the solute per L of solution are turned into moles per L of solution by dividing by the molecular or atomic weight of the solute. Note that you need to make sure your molecular weight is in mg/mol! These in turn are directly related to ppm when the TDS (in mg/L) of the solution is also factored in. In general, since TDS is usually much less than 10000, mg/L is directly convertible to ppm. Only in brines does the TDS start to cause significant (>10%) error.

Chapter 3 Practice Problems

1. If you have a 1000 W microwave, how long does it take to heat a glass of water (250 mL) from room temperature (20°C) to tea-brewing temperature (70°C)? Assume 100% heating efficiency.
2. How many food calories would you burn if you drank 4 L of ice cold (0°C) water per day? How many food calories would you burn if you ate 1 kg of ice (0°C) per day?
3. Will an ice melt if the outside temperature is -1°C and you have added 20 g of NaCl per kg of ice?
4. What is the concentration of Na^+ in mg/L if you add 30 g of NaCl to 500 mL of water? What is its molarity?
5. What is the concentration of K^+ if you add 20 g of K_3PO_4 to 750 mL of water in mg/L? What is its molarity? What is the phosphate concentration in meq/L?
6. A water with a TDS of 100 mg/L has 40 mg/L of Ca^{2+} as a solute. What is this in ppm? In molarity?
7. A brine with a TDS of 250000 mg/L has 5000 mg/L of Na+. What is this in molarity? In ppm?

Chapter 4. Acids and Bases

4.1 What is an acid?

An acid is a chemical compound that, when added to a solvent, acts as an electron receiver within this solvent. In the more geologic case of an acid in water, an acid is a compound that donates a proton to water. Take, for instance, the case of hydrochloric acid:

$$H:\ddot{\underset{..}{Cl}}: + H:\ddot{\underset{..}{O}}:H = :\ddot{\underset{..}{Cl}}:^- + H:\underset{H}{\ddot{\underset{..}{O}}}:H^+$$

The HCl (hydrochloric acid) is acting as both a proton donor and an electron receiver. Both terms describe an acid, but from different perspectives.

4.2 pH

In general, we mostly will consider the chemistry of water with respect to acids and bases. Although acids and bases can be important in some non-aqueous materials, in geology most acid base chemistry will occur in water. For this reason, we often summarize the acidity of a water with a term known as the pH. The pH is the "power" of H^+ dissolved in the water, that is:

$$pH = -\log_{10}[H^+] \qquad (4.1)$$

where the pH is the logarithm of the concentration (in moles/L) of the H^+ ion dissolved in solution. Naturally, the H^+ ion is not free-floating within the water, but is bound to water, forming the hydronium ion (H_3O^+). However, for shorthand, H^+ is fine.

Figure 4.1. pH scale

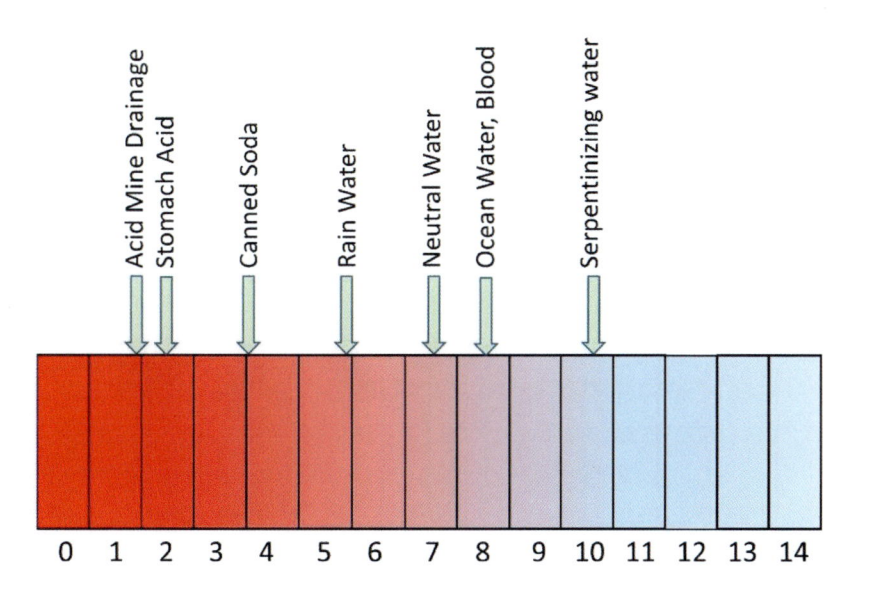

The pH scale provides an estimation of the acidity of a fluid (figure 4.1). Fluids range from extremely acidic (pH of 0 or less) to very basic (pH of 14 or more). Most water falls within a narrower range of pH, typically between 5 and 8. Some waters may be more acidic than this (acid mine drainage may have a pH of 2 or occasionally as low as 1), and some may be more basic (water associated with some hydrothermal systems may become 10-11), but since most water is in contact with a variety of minerals, these waters are buffered by reaction with the minerals to keep them on average around 5-8.

The pH is related to a second term, known as the pOH. The pOH is:

$$pOH = -\log_{10}[OH^-] \quad (4.2)$$

The pOH of a water is used much less frequently than the pH, The pH and pOH are closely related. In a water at a temperature of 298 K (25°C),

$$pH + pOH = 14 \quad (4.3)$$

This comes from something called the K_W of water. At 298 K:

$$[H^+][OH^-] = 10^{-14} \quad (4.4)$$

Take the negative log base ten of both sides to get (4.3).

4.3 Strong acids

There is a suite of acids that historically were the strongest acids known. These acids are listed in table 4.1. These acids are strong enough, that for whatever concentration of acid is added to water, all the acid dissociates to give H$^+$ and a corresponding inert anion (for instance, Cl$^-$). In other words, if you add enough HCl to water to get a 1 M solution of this acid, the acid dissociates and you have a 1 M solution of H$^+$ and of Cl$^-$ with no HCl remaining. There are 6 historically recognized strong acids (and others that have been synthesized in the laboratory, but we'll ignore those). Of these, the only few that are at all commonly found in the environment are H$_2$SO$_4$ that is formed by the oxidation of sulfides and sulfur, HCl that can be produced by reaction of seawater Cl with aerosols and UV light, HNO$_3$ that can be generated by reaction of nitrogen oxide compounds in the atmosphere with water, and HClO$_4$ that may be present on Mars (at least, the ClO$_4^-$ anion appears to be present).

Table 4.1. The strong acids

Name	Formula
Hydrochloric	HCl
Hydrobromic	HBr
Hydroiodic	HI
Nitric	HNO$_3$
Sulfuric	H$_2$SO$_4$
Perchloric	HClO$_4$

4.4 Weak acids, conjugate bases, polyprotic acids

The weak acids (table 4.2) are by far more geologically relevant, and control the chemistry of natural terrestrial fluids to a much greater extent. A weak acid is an acid that, when added to water, does not necessarily fully dissociated. That is, there remains some of the original acid around in solution. The weak acids dissociate to an acidic proton, and to a conjugate base:

$$HA = H^+ + A^- \qquad (4.5)$$

In this example, HA is a generic weak acid, and A- is its corresponding conjugate base. For H$_2$S this is:

$$H_2S = H^+ + HS^-$$

The conjugate base of H$_2$S is HS$^-$. Many weak acids are polyprotic in nature. Polyprotic acids are those acids that possess more than one proton (H$^+$) that can be lost over a pH range of 0-14. On table 4.2, sulfidic acid, phosphoric acid, arsenic and carbonic acids are all polyprotic acids.

Table 4.2. Weak acids

Formula	Name	pK$_A$
H$_2$S	Sulfidic Acid	7.0
H$_2$O$_2$	Hydrogen Peroxide	11.6
H$_3$PO$_4$	Phosphoric Acid	2.1
NH$_4^+$	Ammonium	9.2
H$_3$AsO$_4$	Arsenic Acid	2.3
H$_2$CO$_3$	Carbonic Acid	6.4
HSO$_4^-$	Hydrogen Sulfate	1.9
HF	Hydrofluoric Acid	3.2
H$_3$CCOOH	Acetic Acid	4.7

4.5 The K$_A$

The effectiveness of an acid at losing a H$^+$ is described by its "K$_A$" value. The K$_A$ value is the ratio of products to reactants in the reaction given as (5). That is,

$$\frac{[H^+][A^-]}{[HA]} = K_A \quad (4.6)$$

The K$_A$ is usually unique to a specific weak acid. Strong acids do not have K$_A$s as these dissociate completely and hence the value of [HA] is usually 0. Polyprotic acids have multiple K$_A$s, one for each loss of a proton. The K$_A$s of an acid can be used to calculate the pH of a solution after a weak acid has been added to it, or to calculate the ratio of conjugate base to acid for a given pH. This is also done by using the Henderson-Hasselbalch Equation.

Another way of representing a K$_A$ is as a pK$_A$, Just like the definition of pH, pK$_A$ is given as:

$$pK_A = -logK_A \quad (4.7)$$

Writing pK$_A$s is a shorthand way of avoiding the exponent of K$_A$s.

4.6 The Henderson-Hasselbalch Equation

For a given acid, we can relate the K_A of the acid to the ratio of conjugate base to its corresponding acid. This can be done by taking the negative log base ten of (4.6):

$$-\log \frac{[H^+][A^-]}{[HA]} = -\log K_A$$

$$pH - \log \frac{[A^-]}{[HA]} = pK_A \qquad (4.8)$$

Using this relationship, you can determine the amount of base and acid at any given pH for a weak acid, so long as you know its K_A. The K_A can be determined using thermodynamics.

4.7 Thermodynamics

Thermodynamics is a powerful tool that uses the energy balance and composition of a given chemical system to determine the compounds (species) present for a given temperature and pressure. For this chapter we shall only focus on the thermodynamics of acid-base reactions. For an acid-base reaction, such as:

$$HA = H^+ + A^- \qquad (4.5)$$

The K_A of this reaction can be determined from thermodynamic values. The thermodynamic value most useful for determining compositions in geochemical fluids is the ΔG value, or the Gibbs free energy of a reaction. Thermodynamics uses two equations to do this:

$$\Delta G^0(\text{Reaction}) = \Delta G^0(\text{Products}) - \Delta G^0(\text{Reactants}) \qquad (4.8)$$

$$\Delta G^0(\text{Reaction}) = -RT \ln K$$

or

$$e^{-\Delta G^0(\text{Reaction})/RT} = K \qquad (4.9)$$

For an acid with a reaction in (4.5), the K in (4.9) is the K_A. To do these calculations fully, you need to know the ΔG value for each individual species participating in the chemical reaction.

Chapter 4 Practice Problems

1) What is the pH of a system with 0.01 M H^+? 10^{-5} M H^+? 10^{-3} M OH^-? 3.16 x 10^{-8} M H^+?

2) What is the pH of a 0.01 M solution of H_2S if the K_A of this acid is 10^{-7}?

3) What is the ratio of HSe^- to H_2Se at a pH of 6? The K_A of H_2Se is $10^{-3.9}$.

4) What is the K_A of $HC_2O_4^-$ (oxalate) if the ΔG^0 of H^+ is 0, of $HC_2O_4^-$ is -166.93 kcal/mol, and of $C_2O_4^{2-}$ is -161.1 kcal/mol at 298 K? R is 0.001987 kcal / mol K.

Chapter 5. Thermodynamics

5.1 Principles of thermodynamics

The physical chemistry behind thermodynamics can be a sizeable subject to tackle. I recommend taking physical chemistry if this subject interests you. At this level of course, we shall cover primarily how thermodynamics relates to the ratios of products to reactants in equilibrium systems. This is useful for predicting the pH of solutions, chemical speciation as it relates to redox chemistry, solubility of minerals in fluids, and the extent to which exchange reactions may occur.

At our level of interest, thermodynamics is composed of two terms: ΔH and ΔS. The ΔH term is the change in *enthalpy* of a chemical reaction. This has an obvious relationship to the physical world as this is the heat of a reaction. If a chemical reaction releases or absorbs heat as it proceeds, this is directly linked to the enthalpy of the reaction. If a reaction releases heat, it is called *exothermic*. An example would be the combustion of gasoline. An *endothermic* reaction is the opposite: it absorbs heat to proceed. The evaporation of isopropanol (rubbing alcohol) on your skin is a good example of this process.

The ΔS term is the change in entropy of a chemical reaction. Although entropy is commonly referred to as "disorder", these two terms are not equivalent. Effectively, entropy refers to the ability of a system to explore its physical environment. For instance, if a gas is confined to a vial, if the vial's volume is doubled, the gas can now explore twice as much space and it has increased in entropy. All chemical processes increase the entropy of the universe on some level, though many do so with a smaller loss of entropy elsewhere. The crystallization of salt from water decreases the entropy of the salt (as it goes from freely moving ions to a confined mineral), but this occurs because of evaporation of water (which increases entropy).

These two terms taken together give the Gibbs free energy of a reaction:

$$\Delta G\ (rxn) = \Delta H\ (rxn) - T\ \Delta S\ (rxn) \qquad (5.1)$$

In general, we care primarily about the ΔG (rxn) at standard state (that is, one atmosphere of pressure and 298 K). Under these conditions, ΔG (rxn) is equal to ΔG^0 (rxn).

5.2 The Gibbs Free Energy

The thermodynamic value most useful for determining compositions in geochemical fluids is the ΔG value, or the Gibbs free energy of a reaction. Thermodynamics uses two equations to do this:

$$\Delta G^0(\text{Reaction}) = \Delta G^0(\text{Products}) - \Delta G^0(\text{reactants}) \qquad (4.8)$$

$$\Delta G^0(\text{Reaction}) = - R\,T\,\ln K$$

or

$$e^{-\Delta G^0(\text{Reaction})/RT} = K \qquad (4.9)$$

Note again that R is the universal gas law constant (8.314 J/mol K, or 0.001987 kcal/mol K), and T is the temperature in Kelvin. At standard state, T is equal to 298 K, and the pressure is one atmosphere.

The ΔG^0 values of individual species are compiled as tables of data. These ΔG^0 values are all referenced to the ΔG^0 values of their standard elemental state at a given temperature. In other words, at 298 K, all compounds bearing oxygen are referenced to the thermodynamic properties of O_2 (g). To make it easy, the ΔG^0 of O_2 (g) at 298 K is 0 kcal or 0 kJ or 0 J. Compounds that are composed of multiple elements, for instance Mg_2SiO_4, are referenced to elemental Mg metal, elemental Si, and O_2 gas at 298 K.

The value of a chemical reaction determines the direction of a chemical reaction. If the ΔG (rxn) is greater than 0, then the reaction will move towards reactants, if it's less than 0, then it'll move towards products, and if it's 0 (or close) then it's at equilibrium and won't change much at all. The value of ΔG (rxn) also tells you how much energy can be extracted from a reaction, or whether energy needs to be added to the reaction to push it one direction or another.

5.3 Notes on the K

In most cases we will use thermodynamics to solve for the ratio of products to the reactants, known as the K value. Using equation (3) requires a bit of knowledge on what the K represents. For the generic reaction:

$$aA + bB \rightarrow cC + dD$$

The K value of this reaction is equal to:

$$K = \frac{[C]^c[D]^d}{[A]^a[B]^b} \qquad (5.2)$$

Note that each of the coefficients in front of the compounds reacting become exponents in the K. The value used for [C] is the *activity* of C. The activity of C is unitless and represents a reactive characteristic of the element. In most of the cases we will explore, the activity will be set equal to the molarity of C (the moles of C per liter of solution). This is accurate for most water samples except for those with a high TDS.

There are some other curiosities associated with activity. Most solids are defined in this system as having an activity of 1. In other words, solids do not participate in K calculations (as a ratio multiplied or divided by 1 is equal to itself). Naturally this is not always true, but serves as a useful approximation. Additionally, the activity of water will usually have a value of 1 as well, so, like solids, it drops out of the K calculation. Finally, if a gas participates in a chemical reaction, its activity (called its fugacity) can be calculated from Raoult's and Henry's law as:

$$f_X = X_X P_{Total} \quad (5.3)$$

where f_X is the fugacity of gas X, X_X is the mole fraction of gas X (which is typically equivalent to the volume fraction of gas X in the atmosphere), multiplied by P_{Total}, or the total pressure of the gas above the solution.

5.4 Non-equilibrium systems

For the most part these calculations are used for equilibrium systems. For a non-equilibrium system, which is one that has not completely reached the lowest energy, equation (3) is modified to:

$$\Delta G = \Delta G^0 + RT \ln Q \quad (5.4)$$

Where ΔG^0 is the Gibbs free energy at equilibrium, and Q is the reaction quotient at the disequilibrium state. Note that the ΔG being calculated does not need to be at standard state, and is not necessarily at equilibrium. Indeed, at equilibrium the ΔG value is equal to 0, and there is no chemical energy free to be extracted from the reaction.

Q is equivalent to K, but has the values that are measured at a given point in the reaction placed into the activities instead of the equilibrium values.

5.5 The Van't Hoff equation

The thermodynamics we have discussed so far have been useful only around 298 K. The calculated values are only exact at precisely 298 K, though most K values don't tend to vary too much within about a five degree range of 298 (or vary at most by a factor of 2). We can modify the thermodynamics using the Van't Hoff equation to calculate K values over a much large temperature range:

$$\ln \frac{K_2}{K_1} = \frac{-\Delta H^0_{rxn}}{R} \left(\frac{1}{T_2} - \frac{1}{T_1} \right) \quad (5.5)$$

Where K_2 is the K at temperature T_2, K_1 is the K at temperature T_1, ΔH^0_{rxn} is the calculated enthalpy of the reaction (found in a process identical to equation 4.8), and R is the gas law constant. In general, you would solve first for the K at 298, and use this as your K_1 and T_1 values, then solve for a different K at a different given T_2. This methodology is valid over a much larger temperature range (about 100 degrees for many systems, but only if there are no phases changes).

Chapter 5 Practice Problems

The procedure for using thermodynamics to solve for chemical system compositions is to first write a balanced chemical reaction, then to solve for the ΔG^0 (rxn), then to solve for the K, and then to use this K to address your questions.

1) What is the concentration of Na^+ in a water sample that is saturated with NaCl?

2) What is the maximum concentration of silica in water in a water that is in direct contact with quartz?

3) At what pH does H_2S coexist with HS^- in equal abundance?

4) How much O_2 does surface water contain?

5) What is the pH of a glass of water?

6) Can iron metal reduce Phosphate?

$$Fe + HPO_4^{2-} \rightarrow FeO + HPO_3^{2-}$$

Chapter 6. Redox Reactions

6.1 Determining oxidation state

Understanding the oxidation state of an element is a critical part of understanding the chemistry of a solution, including natural waters. Water from different locations can be *reducing* or *oxidizing*, which tells you something about the chemistry of the water.

An important first step in all oxidation-reduction chemistry (also known as *redox* reactions) is determining the oxidation state of an element. The oxidation state is related to the number of extra electrons or to the electron deficiency an element has. In general, the oxidation state should be viewed as a tool to this end, as the exact electron distribution in a molecule is not always as clear cut as this process would state.

Determining the oxidation state of an element uses the following rules:

- All native elements have an oxidation state of zero. For instance, S, S_8, O_2, and Fe metal are all oxidation state 0.
- The charge on a dissolved monoatomic ion is its oxidation state. For instance, dissolved Fe^{2+} or Mg^{2+} both have oxidation states of +2. Al^{3+} (aq) has an oxidation state of +3.
- All alkali metals have an oxidation state of +1 if they are not in metallic form. Na^+ (aq), Li^+ (aq), and the Na in Na_2O all have an oxidation state of +1.
- All alkali earth metals have an oxidation state of +2 if they are not in metallic form. Ca^{2+} (aq) and the Mg in MgO all have an oxidation state of +2.
- In general, most elements follow the trends of periodicity. For instance, the boron group elements (B, Al, Ga, In, Tl) tend to have oxidation +3. The carbon column are mostly +4. The column with N tend to be -3 (but not always), with O are usually -2 (but not always), and the halogens are usually -1 (but not always!). Noble gases are almost always 0. Transition metals are messier, with oxidation states between 0 and +8 in most cases.
- As a good rule, oxygen has an oxidation state of -2 except when it occurs as O_2 gas (oxidation state 0), or as hydrogen peroxide H_2O_2, with an oxidation state of nominally -1.
- Hydrogen tends to have an oxidation state of +1, except as hydrogen gas (H_2) with an oxidation state of zero.
- For a dissolve polyatomic ion, the sum of the oxidation states of individual elements is equal to the total charge of the ion. That is, $C_{Total} = \Sigma\, C_{individuals}$

Naturally, as with most science, there are plenty of exceptions to these rules. However, in the realm of geochemistry, these exceptions are not common.

Example: determine the oxidation state of As in $H_2AsO_3^-$. Since you know the total charge of the ion is -1, this tells you that the sum of the oxidation states of all of the O's, H's, and the As must be equal to -1. Since each H has a +1 charge, and each O has a -2 charge, you can calculate this as follows:

$C_{Total} = \Sigma\ C_{individuals} = -1 = 2 \times C_H + C_{As} + 3 \times C_O = 2 \times (+1) + C_{As} + 3 \times (-2);$ $C_{As} = +3$

You can use this method to determine the charge of P in HPO_3^{2-} (it's +3), and of P in PH_3 (it's -3).

6.2 Volts

The act of changing the oxidation state of an element occurs through a redox reaction. For an element to change its redox state, something else must change its redox state in the opposite direction to balance the flow of electrons. In redox reactions, for something to lose electrons (become oxidized) something else must gain the electrons (become reduced). This can be remembered by the mnemonics: LEO the lion says GER (Loses Electrons, Oxidized; Gains Electrons, Reduced), or OIL RIG (Oxidation Is Loss of electrons, Reduction Is Gain of electrons). Alternatively, the term oxidation comes from the increase in oxygen numbers that usually corresponds to an oxidation process (for instance, H_3PO_3 to H_3PO_4) whereas reduction is a loss of oxidation number (for instance, from H_2SO_4 to H_2S), which corresponds to a reduction in mass.

The direction of a coupled redox reaction is strongly linked to the free energy of a reaction. It is also important in everyday processes, as the shuffling of electrons is denoted in energy terms by Volts. In the generic reaction:

$$A\ +\ B\ \rightarrow\ A^+\ +\ B^-$$

A must transfer an electron to B. The transfer of this electron (which requires energy to pluck the electron from A, and gives energy to put the electron onto B, generally) has an energy potential associated with it. That is, each electron that may participate in this reaction has a potential energy associated with it (either positive or negative).

Naturally, the shuffling of electrons is closely related to thermodynamics. In the general case:

$$\Delta G = n\ F\ E \qquad (6.1)$$

That is, the free energy of the reaction (ΔG) is equal to equal to the number of electrons (**n**) that shuffle about in a balanced (see section 4 below) chemical reaction, multiplied by a constant called the Faraday constant (F), multiplied by the energy potential or voltage (E). A voltage has units of volts, which are joules per coulomb. A coulomb is a count of electrons, much like moles. More specifically, there are 96485 coulombs in a mole of electrons. The Faraday constant has a value of 96485 if you are turning volts into a ΔG with units of J/mole, or 23054 cal/mole, or 23.054 kcal/mole.

6.3 Redox reaction half cells

An important part of understanding redox reaction is determining the flow of electrons in those reactions. In a redox reaction, the *oxidizing agent* is the thing that is reduced (gains electrons), and the *reducing agent* is the thing that is oxidized (loses electrons).

A redox reaction can be broken into two parts: the thing that is oxidized, and the thing that is reduced. For instance, you can rewrite:

$$A + B \rightarrow A^+ + B^-$$

As:

$$A \rightarrow A^+ + e^-$$
$$B + e^- \rightarrow B^-$$

These are called half-cell reactions, and show better how electrons flow. The utility of this is twofold. For one, you can get thermodynamic data easily from half-cells as the energy potential for many redox reactions is well known. For instance, the voltage of calcium oxidation is:

$$Ca \rightarrow Ca^{2+} + 2e^- \qquad -2.87\ V$$

The corresponding **ΔG** is hence calculated from equation (1) as -132.3 kcal/mol (-553.8 kJ/mol). Additionally, half cells allow for the easy balancing of electrons. For instance, how platinum and indium react:

$$Pt + In^{3+} \rightarrow In + Pt^{2+}$$

Can be determined by breaking this into half cells

$$Pt \rightarrow Pt^{2+} + 2e^-$$
$$In^{3+} + 3e^- \rightarrow In$$

Since Pt loses 2 electrons and Indium gains 3, and no electrons should be in the final reaction, multiply the platinum reaction by 3 and the indium reaction by 2 to get a balanced redox reaction:

$$3Pt + 2In^{3+} \rightarrow 2In + 3Pt^{2+}$$

6.4 Balancing redox reactions

Balancing a redox reaction can be tricky. It is more difficult than simply balancing the elements of a reaction (e.g., $A_2B_3 \rightarrow 2A^{3+} + 3B^{2-}$) as there is a requirement to balance electron flow. This is done by following these five steps (the **redox algorithm**):

1. Balance the non-O, and non-H elements that are participating in the reaction
2. Break the reaction into half cells, one with the element being reduced, the other with the element being oxidized. Determine how many electrons are being gained and lost, respectively. Multiply these half-cell reactions by a number required to make the number of electrons lost equal to the number of electrons gained. Add the two reactions together.
3. Balance oxygen on both sides of the reaction with water (H_2O).
4. Balance charge on both sides of the reaction with H^+.
5. Check to see that you've done it correctly by counting the number of hydrogens on both sides of the reaction.
6. There is a special case for O_2 reacting. In some reactions, O_2 will become oxidized. In this case, O_2 turns into "2 O^{2-}" by gaining 4 electrons. If so, one of the half-cells will be precisely this: $O_2 + 4e^- \rightarrow 2O^{2-}$. The O^{2-} is not counted in the redox reaction when the two half-cells are joined (and disappears, effectively).

There are many ways to balance a redox reaction; the above is merely a functional guideline. Fortunately, independent of how you balance a reaction, you will get the same answer with respect to thermodynamics and environmental conditions (if done properly!).

An example of this process would be illustrative. Consider:

$HMnO_4^- + H_2S \rightarrow MnO + S_8$

Step 1, balance the non-O and H:

$HMnO_4^- + \mathbf{8}H_2S \rightarrow MnO + S_8$

Step 2, break these two reactions into half-cells:

$HMnO_4^- \rightarrow MnO$

$8H_2S \rightarrow S_8$

For the first one, the oxidation state of Mn in $HMnO_4^-$ is +6, and in MnO is +2, thus Mn has gained four electrons. In the second, the charge on S in H_2S is -2, and in S_8 is 0, thus each S has lost two electrons. Since there are eight sulfurs each losing 2 electrons, a total of 16 are lost.

$HMnO_4^- + 4e^- \rightarrow MnO_2$

$8H_2S \rightarrow S_8 + 16e^-$

The electrons within these two reactions can be balanced by multiplying the top (Mn) one by 4 and keeping the bottom one the same.

$$4HMnO_4^- + 16e^- \rightarrow 4MnO_2$$

$$8H_2S \rightarrow S_8 + 16e^-$$

When added together you get:

$$4HMnO_4^- + 8H_2S \rightarrow 4MnO + S_8$$

Step 3, balance your O with H_2O. The reactants have a total of 4 × 4 = 16 O, the products have 4 O's. Thus the products need 12 more O's as H_2O to be balanced:

$$4HMnO_4^- + 8H_2S \rightarrow 4MnO + S_8 + 12H_2O$$

Step 4 is to balance the charge with H^+. The total charge on the reactant side is -4 (4 × -1 + 8 × 0), whereas on the product side the total charge is 0 (no anions or cations whatsoever). Thus the reactants need 4 H^+ to balance charge:

$$4HMnO_4^- + 8H_2S + 4H^+ \rightarrow 4MnO + S_8 + 12H_2O$$

We have finished the balancing, and **step 5** is to check our numbers. On the reactant side we have 4 Mn, 16 O, 8 S, and 24 H. On the product side we have 4 Mn, 16 O, 8 S, and 24 H. Thus the reaction is balanced.

6.5 Solving for redox using thermodynamics

Once you have a balanced redox reaction, you can calculate the voltage of the reaction by using relationship (6.1). So in the case of the previously solved redox reaction:

$$4HMnO_4^- + 8H_2S + 4H^+ \rightarrow 4MnO + S_8 + 12H_2O$$

You can solve for the ΔG^0 of this reaction and then the E^0 of this reaction using $\Delta G^0 = n F E^0$. Do so as a typical thermodynamics calculation:

ΔG^0(Reaction) = ΔG^0(Products) – ΔG^0(reactants)

=(4 × ΔG^0 MnO + 1 × ΔG^0 S_8 + 12 × ΔG^0 H_2O) – (4 × ΔG^0 $HMnO_4^-$ + 8 × ΔG^0 H_2S + 4 × ΔG^0 H^+)

= (4 × -86.74 + 0 + 12 × -56.687) – (-118.1 + 6 × -6.66 + 4 x 0) = -869.144 kcal

Since there are 23.054 V per kcal, and there are 16 electrons participating in this reaction, you can calculate the E^0 from these two numbers.

$E^0 = \Delta G^0 / nF = -869.144 / (16 \times 23.054) = -2.35$ V

Most E^0 values should be between -3 and +3 V.

6.6 The Nernst Equation

Recall that for non-equilibrium thermodynamic calculations, ΔG can be found from chapter 5, section 4:

$$\Delta G = \Delta G^0 + RT \ln Q$$

Where ΔG^0 is the Gibbs free energy at equilibrium, and Q is the reaction quotient at the disequilibrium state. You can combine this with relationship (1) by dividing everything by nF. This gets you:

$$\frac{\Delta G}{nF} = \frac{\Delta G^0}{nF} + \frac{RT}{nF} \ln Q \qquad (6.2)$$

Note that this becomes:

$$E = E^0 + \frac{RT}{nF} \ln Q \qquad (6.3)$$

In general, for most redox reactions you're only going to be considering one temperature (298 K), and you will tend to like to do things in log base ten. Recall that changing bases is done through:

$$\ln Q = \text{Log}_{10} Q / \text{Log}_{10} e$$

Thus the RT/nF term shrinks down to:

$$E = E^0 + \frac{0.0592}{n} \log Q \qquad (6.4)$$

This is known as the Nernst equation, and is extremely useful in estimating redox reactions and conditions. It features prominently in establishing lines on an Eh-pH diagram, a diagram of the speciation of materials as functions of oxidation-reduction potential, and pH.

Chapter 6 Practice Problems

1) Determine the oxidation state of P in $H_2PO_2^-$, of Fe in $Fe(OH)_2^+$ and $FeOH^{2+}$, and of Al in $Al(OH)_4^-$.

2) Balance the following reactions. Calculate their E^0

A) Ag^+ + Cu → Ag + Cu^{2+}
B) Cl_2 + HCO_3^- → ClO_4^- + CH_4
C) PbO_4^{4-} + N_2O → Pb^{2+} + NO_3^-

Chapter 7. Eh-pH diagrams

7.1 The Nernst Equation

Since imitation is the sincerest form of flattery, let's repeat. Recall that for non-equilibrium thermodynamic calculations, ΔG can be found from

$$\Delta G = \Delta G^0 + RT \ln Q \qquad (5.4)$$

Where ΔG^0 is the Gibbs free energy at equilibrium, and Q is the reaction quotient at the disequilibrium state. You can combine this with relationship (1) by dividing everything by nF. This gets you:

$$\frac{\Delta G}{nF} = \frac{\Delta G^0}{nF} + \frac{RT}{nF} \ln Q \qquad (6.2)$$

Note that this becomes:

$$E = E^0 + \frac{RT}{nF} \ln Q \qquad (6.3)$$

In general, for most redox reactions you're only going to be considering one temperature (298 K), and you will tend to like to do things in log base ten. Recall that changing bases is done through:

$$\ln Q = \text{Log}_{10} Q / \text{Log}_{10} e$$

Thus the RT/nF term shrinks down to:

$$E = E^0 + \frac{0.0592}{n} \log Q \qquad (6.4)$$

This is known as the Nernst equation, and is extremely useful in estimating redox reactions and conditions. It features prominently in establishing lines on an Eh-pH diagram, a diagram of the speciation of materials as functions of oxidation-reduction potential, and pH.

Redox reactions that occur in water are referenced to a redox reaction called the standard hydrogen electrode. This electrode, which is based on the chemical reaction:

$$H_2 \rightarrow 2 H^+ + 2 e^-$$

Is the one to which all other redox reactions occurring in water are referenced. This allows us to change the Nernst equation to

$$Eh = E^0 + \frac{0.0592}{n} \log Q$$

Where the Eh means that the redox potential is referenced to the standard hydrogen electrode.

7.2 Reactions within an Eh-pH diagram

Recall that I introduced last chapter the concept of a half cell – a chemical reaction in which a single material is either reduced or oxidized. Of course, you can rewrite any reduction reaction to represent an oxidation reaction just by exchanging the products with the reactants. For most aqueous geochemical systems, we deal primarily with two terms to understand the speciation of dissolved species in water: redox and pH. A chemical reaction in this phase space may be one of three types: it may be an acid-base reaction only, it may be a redox reaction only, or it may be both a redox and an acid base reaction.

An acid-base reaction is one that only consists of exchange of H^+ or OH^-, and the given element has no change of oxidation state. Not all acid-base reactions are obvious. Consider, for instance, the acid-base chemistry of aluminum:

$$Al^{3+} + H_2O \rightarrow AlOH^{2+} + H^+$$

This is a balanced reaction (same numbers of elements on both sides), and aluminum remains the same oxidation state, however, the reaction results in the kicking out of a proton, which by definition is an acid-base reaction. Thus Al^{3+} is an acid.

Consider also the oxidation of iron:

$$Fe^{2+} \rightarrow Fe^{3+} + e^-$$

This reaction is balanced (same numbers of irons on both sides, same charge on both sides), but there is no participating H^+. This is a redox-only reaction.

Consider as a final case the oxidation of H_2S to SO_4^{2-}:

$$H_2S + 4\,H_2O \rightarrow SO_4^{2-} + 10H^+ + 8\,e^-$$

In this redox reaction, you see dependences on both the acid-base chemistry, and on the electron flow. If we were to put these reactions into the Nernst equation, we would get the following:

$$Eh = E^0 + \frac{0.0592}{0} \log \frac{[AlOH^{2+}][H^+]}{[Al^{3+}]}$$

$$Eh = E^0 + \frac{0.0592}{1} \log \frac{[Fe^{3+}]}{[Fe^{2+}]}$$

$$Eh = E^0 + \frac{0.0592}{8} \log \frac{[SO_4^{2-}][H^+]^{10}}{[H_2S]}$$

Note that the electrons cancel out, and that water has an activity of 1. For these reactions, let's consider the region where the reduced form of the element is as abundant as the oxidized form of the element (e.g., $[Fe^{2+}]=[Fe^{3+}]$). At these points, these species cancel out and you are left with equations that can be summarized as a line:

$$Eh = E^0 - \frac{0.0592}{n} A \times pH$$

$$Y \;\;=\;\; b \;-\; m\;(X)$$

Where A is the number of H^+ in the balanced redox reaction, and n is the number of electrons. An Eh-pH diagram traces the relationship between pH (x-axis) and Eh (y-axis). For reactions where there are no electrons (e.g., the aluminum one above) the slope becomes 0.0592 / 0, or effectively infinite (vertical in Eh-pH space). In contrast, if a reaction has no H^+ to balance it, then A = 0, hence the slope is 0. If the reaction has both electron transfer and acid-base chemistry, the slope is not 0 or infinity and is usually negative.

7.3 Boundaries of Eh-pH

Although all redox reactions might theoretically be possible under chemical environments, for the average natural terrestrial environment, the extent of redox variability is minimal. Two lines define the boundaries of Eh-pH, and these lines are usually represented as dashed lines on the Eh-pH diagram. The bottom line is the reducing limit of the stability of water. Below this line, the hydrogen in water is reduced to H_2. The upper line is the oxidizing limit of the stability of water. Above this line, the oxygen in water is oxidized to O_2.

Most Eh-pH diagrams are also constructed with a critical assumption: the maximum dissolved concentration of the elements being investigated is a set number (usually 10^{-4} M, 10^{-6} M, or 10^{-8} M). This allows for easier cancellation of species during the preparation of the Nernst equation.

7.4 The Eh-pH diagram

The only real way to become familiar with Eh-pH diagrams is to see them. Four diagrams are outlined below. These are all modified from Takeno (2005), the Atlas of Eh-pH diagrams.

Aluminum is a fairly boring element. The only lines you see on its Eh-pH diagram are acid-base transitions. In general, aluminum is only found in the natural environment as Al^{3+}.

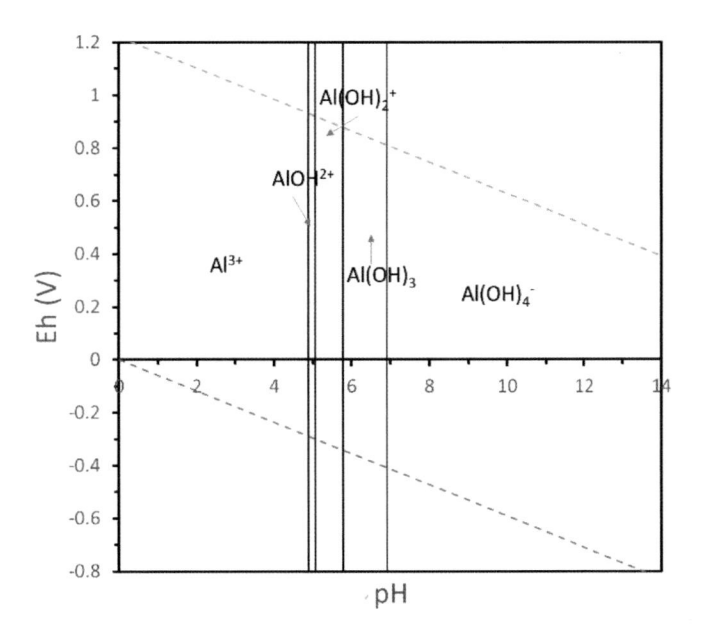

Silver is a bit more interesting. However, most of the chemistry of silver is a redox transition between Ag and Ag$^+$. There is also some acid-base chemistry for Ag$^+$ at a pH of about 12.

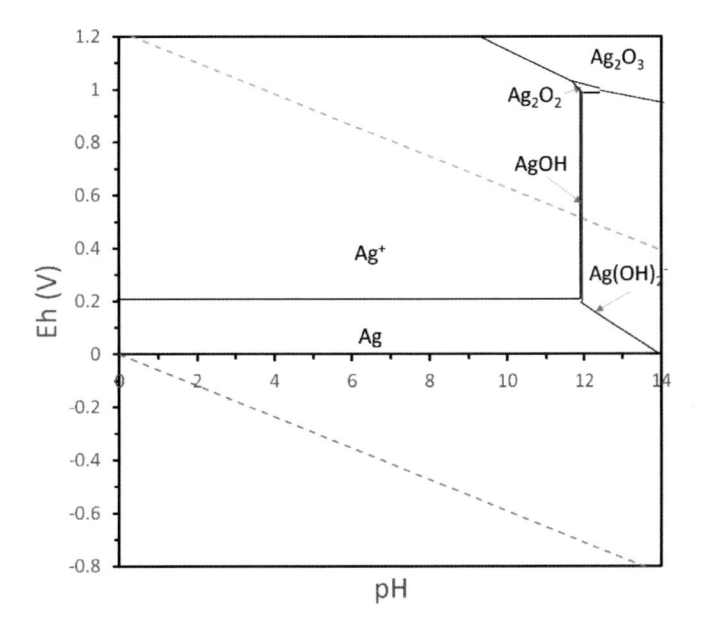

Nitrogen only has one acid-base transition for ammonium/ammonia. The rest are Eh-pH transitions, and there are 3 major redox states for the element (-3 ammonia, 0 N$_2$, and +5 nitrate).

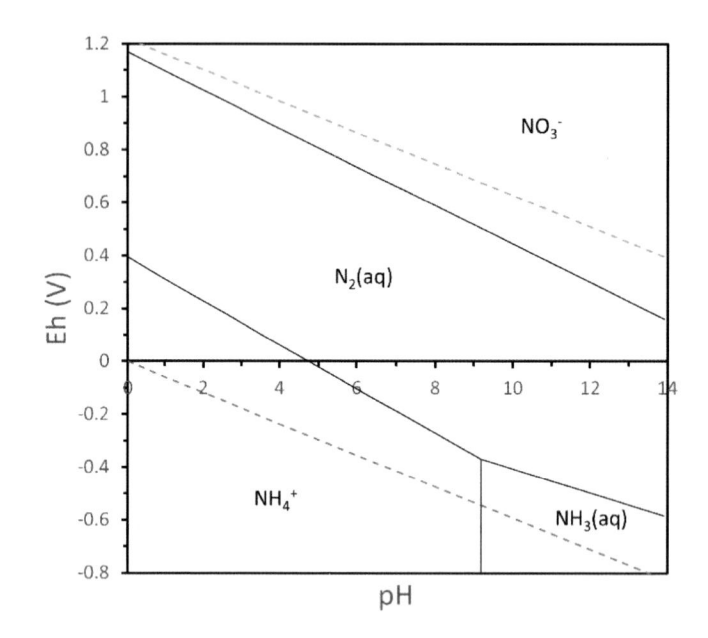

Plutonium has a lot of species and reactions.

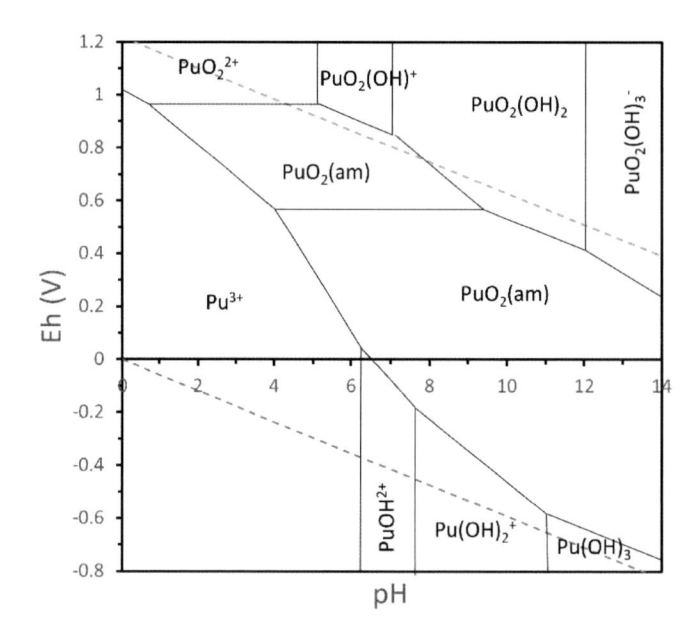

Chapter 7 Practice Problems

1. Using the Eh-pH diagram for Sulfur, identify the major redox states present, and the acid-base transitions that are taking place within each redox state.

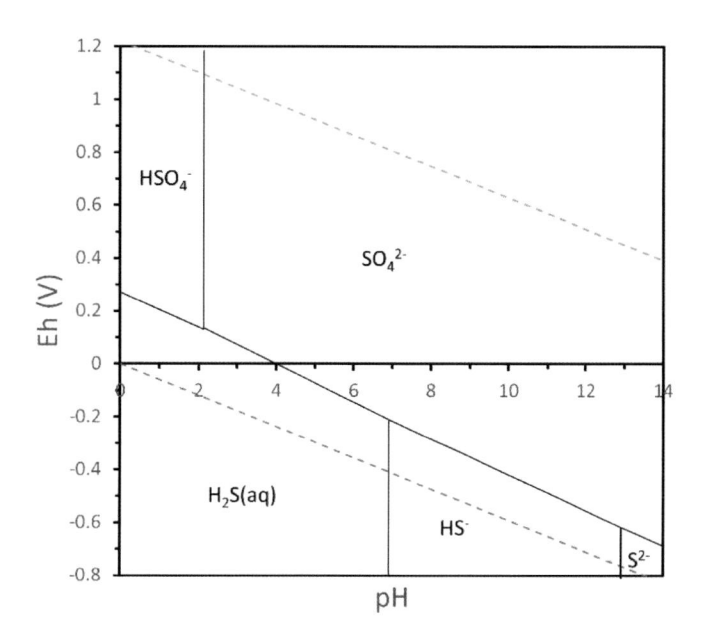

2. What solids of neptunium might you expect in a natural groundwater?

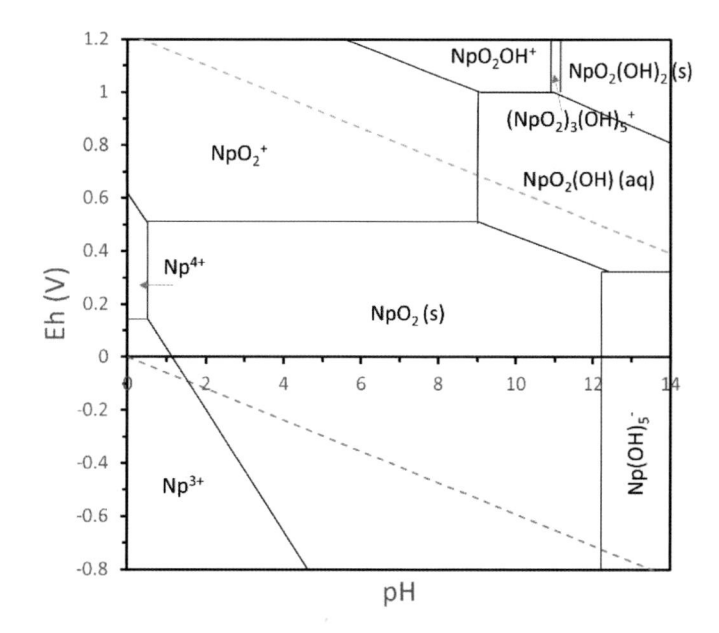

3. What conditions are conducive to methanogenesis?

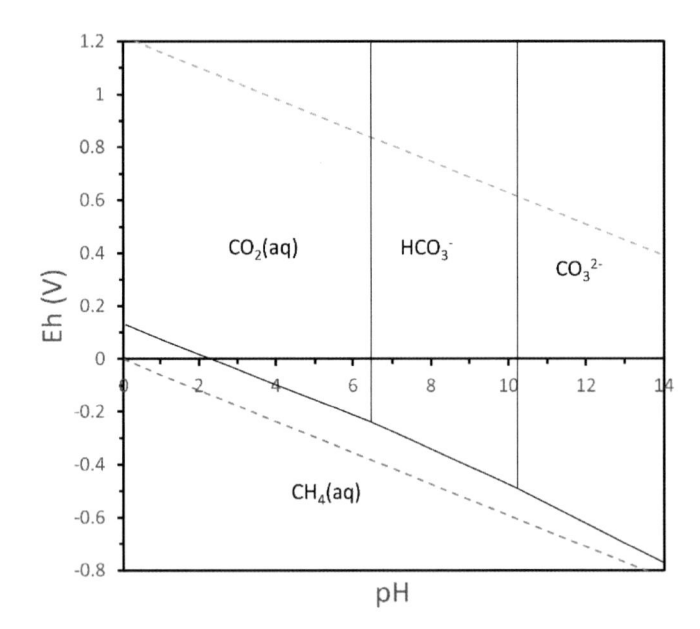

Chapter 8. Radioactivity

8.1 Derivation of the first order decay process

Time for a bit of calculus! Let's assume that a substance (A) will disappear with time. The rate of its disappearance is directly proportional to its own abundance, that is:

$$\frac{dA}{dt} = -kA$$

where k is a constant, but not equal to the K we discussed in thermodynamics. Using this relationship, we can rearrange this to:

$$\frac{dA}{A} = -kdt$$

We can then integrate this from a starting time (time zero or t_0) to a time of interest (time t or t):

$$\int_{t_0}^{t} \frac{dA}{A} = -k \int_{t_0}^{t} dt$$

This becomes:

$$\ln \frac{A}{A_0} = -kt$$

Which rearranges to give:

$$A = A_0 e^{-kt} \qquad (8.1)$$

This is the first order decay process. Let's also assume that there is some sort of characteristic timescale over which half of the initial amount of A present (A_0) has decayed. We'll call this the half-life ($t_{1/2}$). This half-life is equal to:

$$\ln \frac{A}{A_0} = \ln \frac{1}{2} = -kt_{1/2}$$

The half-life is thus equal to:

$$\frac{\ln 2}{k} = t_{1/2} \qquad (8.2)$$

The half-life is a useful number that is characteristic of a single radioactive decay process. If you start with 1000 atoms, over the course of one half-life you will have 500 left. After 2 half-lives, you have 250, and 125 after 3 half-lives, and so on.

8.2 Radioactive decay modes

Radioactivity is useful in geology for one major reason: it is unaffected by chemistry. Radioactivity is almost completely a nuclear process; it occurs primarily within the nucleus of an atom. Since the nucleus of an atom is unaffected by chemistry (recall that chemistry is primarily affected by electrons), the rate of decay of a radioactive atom is also unaffected by chemistry. For instance, potassium-40 is a common radioactive atom. Potassium-40 has a half-life of 1.25×10^9 years (what is its k?- see answer below).

Radioactivity is the process by which a high-energy nucleus becomes a lower energy nucleus. This is accomplished by four major decay modes. In these examples, we denote an isotope as the following:

$$^N_A X:$$

where X is the element, A is its atomic number (or number of protons) and N is its atomic weight (equal to number of protons plus the number of neutrons).

Alpha Decay: In alpha decay, a radioactive nucleus ejects a helium nucleus, losing 2 protons and 2 neutrons. The net effect is a drop in weight by 4, and of atomic number of 2.

$$^{238}_{92}U \rightarrow ^{234}_{90}Th + ^4_2\alpha$$

The ejected helium nucleus is called an alpha particle. This nuclear decay process occurs primarily to heavy nuclei (things weighing more than nickel, and generally within the lanthanide elements or heavier).

Beta Decay: In beta decay, a radioactive nucleus ejects an electron, turning a neutron into a proton. There is no net change in mass, but an increase in atomic number.

$$^{98}_{43}Tc \rightarrow ^{98}_{44}Ru + ^0_{-1}\beta$$

The ejected electron is called a beta particle. This nuclear decay process occurs primarily to elements that have an excess of neutrons relative to protons. It often occurs in concert with alpha decay. In most elements lower in weight than calcium, the number of protons is almost equal to the number of neutrons. In elements heavier than calcium, there are more neutrons than protons (about 1.2 neutrons per proton). Thus, if an atom has too many neutrons relative to its protons (via alpha decay), beta decay allows the atom to redistribute its nuclear particles to maintain parity better.

Positron Emission: In positron emission, a radioactive nucleus ejects a positron, turning a proton into a neutron. There is no net change in mass, but a decrease in atomic number.

$$^{18}_{9}F \rightarrow\ ^{18}_{8}O + ^{0}_{+1}\beta$$

This nuclear decay process occurs primarily to elements that have an excess of protons relative to neutrons. It is easily detected by simple radiation meters as the positron rapidly annihilates an electron (as it is an antimatter electron), releasing energy proportional to its mass (e.g., $E = mc^2$).

Electron Capture: In electron capture, a nucleus absorbs an inner shell electron, turning a proton into a neutron. There is no net change in mass, but a decrease in atomic number.

$$^{197}_{80}Hg + ^{0}_{-1}e \rightarrow\ ^{197}_{79}Au$$

This nuclear decay process occurs primarily to elements that have an excess of protons relative to neutrons. It is one of the few radioactive decay processes that can be affected by chemistry, on an extremely small scale and only under extreme conditions (such as the interior of stars).

Answer: The k of Potassium-40 is equal to $(\ln 2) / 1.25 \times 10^9$, or 5.55×10^{-10} yr^{-1}

8.3 Important equations

Uses of radioactivity in geology center around three equations:

$$A = A_0 e^{-kt} \quad (8.1) \quad \text{and} \quad A_0 = A + D^* \quad (8.3) \quad \text{and} \quad D = D^* + D_i \quad (8.4)$$

In these equations, A is the amount of radioactive nuclide at time t, A_0 is the amount at the beginning (t = 0), D is the amount of daughter atom present in the material being analyzed, D^* is the amount of daughter formed by radioactive decay, and D_i is the amount of daughter initially present in the material (not from the radioactive decay being studied). These equations mean that you must know A and one of the following: A_0, D^* or D and D_i to get an age. From these equations, several other relationships can be determined, and many of these are useful depending on the isotopic situation of interest.

8.4 Radiometric Decay Methods

An assortment of radiometric decay dating methods are available for geologists (table 8.1). These can be employed to varying degrees of success depending on expected age and the material being sampled. Some decay methods are only valid for age dating young material, whereas others are only useful for age dating older objects (millions to billions of years in age). Most age-dating methods are only useful for about 10 half-lives of time.

Table 8.1. Common radiometric dating techniques and their half-lives

Method	Half-life (y)
Radiocarbon	5730
^{235}U-^{207}Pb	7.04 x 10^8
^{238}U-^{206}Pb	4.47 x 10^9
K-Ar	1.25 x 10^9
Rb-Sr	4.88 x 10^{10}
^{36}Cl	301,000
T	12.26

Radiocarbon. Carbon-14 is one of the oldest methods of dating samples, and has found extensive use in archeology. Carbon 14 is generated in the atmosphere by cosmic ray bombardment of nitrogen 14. All organisms that absorb carbon from the atmosphere (e.g., land plants) or that eat these organisms maintain a low-level concentration of ^{14}C. Once an organism dies, this renewal of ^{14}C stops and the radiocarbon begins to decay. Equation (1) is useful for radiocarbon dating.

Uranium-Lead. This method is useful for uranium-rich, lead-poor minerals (such as zircons), and the rocks that bear these mineral, including granites. This method usually measures the uranium and lead isotopes of interest as ratios to a stable isotope, typically ^{204}Pb, as ratios are easier to measure by mass spectrometry. Both ^{238}U and ^{235}U and their respective lead daughter atoms are measured to determine if the dates are concordant (have the same age). Many rocks show concordance in their dates. However, rocks that deviate from concordance can be plotted on a Concordia diagram to determine the age of formation (oldest age) and the age of the last significant metamorphic event (youngest age). This is determined by drawing a straight line through the data. Since you typically measure the ratio of lead daughter to uranium parent atoms, equations (1) and (2) can be combined to get:

$$\frac{D^*}{A} + 1 = e^{kt} \qquad (8.5).$$

If the amount of radioactive daughter produced is not known, then a variation on this equation should be used:

$$A(e^{kt} - 1) + D_i = D \quad (8.6)$$

Rubidium-Strontium. Use of Rb-Sr dating requires several materials of differing Rb content. These can be different minerals, different rocks associated with the same formation, or sometimes different rocks altogether. This is because most rocks contain a lot of strontium, which is the daughter product of ^{87}Rb. This method usually measures the rubidium-87 and strontium-87 isotopes as ratios to the stable isotope ^{86}Sr, as ratios are easier to measure by mass spectrometry. Since most rocks contain a lot of strontium, the date of these rocks is determined by the isochron method. This is done by employing equations (8.1), (8.3), and (8.4) to get a relationship:

$$A(e^{kt} - 1) + D_i = D \qquad (8.6).$$

Multiple values of A (Rb-87) and D (Sr-87) are determined, and plotted as a line. The slope of the line is equal to e^{kt}-1, allowing a determination of age.

Potassium-Argon. This method is used on rocks with abundant potassium. Since potassium and argon are the most mobile elements of the solid rock radiometric dating techniques, these elements are the most easily affected by metamorphism and heating. Age dating with K-Ar primarily uses equation (8.5) with a caveat. Argon-40 is produced only 10.9% of the time from the decay of K-40. The other 89.1% of the time, Ca-40 is produced. Thus arriving at a correct D* requires factoring this in. Dividing the amount of ^{40}Ar by 0.109 corrects for this.

Tritium and other cosmogenic radionuclides. An assortment of other dating techniques may be employed, depending on the desired material to be measured. Tritium, radiocarbon, and ^{36}Cl have all been used in groundwater studies. In these studies, it is assumed that there is some sort of steady state production of these isotopes, for instance by cosmic ray interactions with nitrogen (for radiocarbon) or argon (for ^{36}Cl). Most of these dating techniques use variations of equation (8.1).

Chapter 8 Practice Problems

1) A mummy bears a radiocarbon signal equal to 60% of the present-day value. How old is it?
2) A zircon is found to bear 0.3 ppm ^{207}Pb and 10 ppm ^{235}U. How old is it?
3) A zircon has a ^{206}Pb / ^{204}Pb = 59.7 and ^{238}U / ^{204}Pb = 22.1. Assume the initial ^{206}Pb / ^{204}Pb in the zircon was 40.2. How old is it?
4) Several meteorites have the following Rb/Sr ratios:

Meteorites	^{87}Rb/^{86}Sr	^{87}Sr/^{86}Sr
Modoc	0.86	0.757
Homestead	0.8	0.751
Bruderheim	0.72	0.747
Kyushu	0.6	0.739
Buth Furnace	0.09	0.706

How old is the solar system?

5) A lunar breccia has ^{40}Ar = 9.57 ppm, ^{40}K = 12.16 ppm. How old is it?
6) A bottle of wine has half of the present-day tritium concentration in it. How old is it?
7) A water 300 km from its source has a ^{36}Cl concentration equal to 22% of the present-day concentration. What is the flow rate of the water?

Chapter 9. Stable Isotopes

9.1 Isotopes

Recall that an element is defined by its number of protons. The number of protons an element possesses determines the element, that is, an atom with 75 protons is a rhenium atom, independent of the number of electrons and neutrons it has. The chemistry of rhenium is mostly defined by the number of electrons and protons an atom has, for instance, if rhenium has only 73 electrons, then its redox state would be +2 and it would behave as a divalent cation of a specific size. In general, the number of neutrons an atom has does not significantly influence its chemistry, beyond determining whether the atom is stable or unstable and decays by radioactive decay.

In general, an isotope is denoted as $^N_A X$: where X is the element, A is its atomic number (or number of protons) and N is its atomic weight (equal to number of protons plus the number of neutrons). A is often left off as A can also be determined by the element.

Many elements possess multiple isotopes that are stable. These isotopes vary in the number of neutrons they have, but this variation is not enough to cause an isotope to become unstable. Examples of stable isotopes are listed in table 9.1. There are over 200 stable isotopes known, and several radioactive isotopes have suitably long half-lives to be considered to be stable over a short geologic timescale.

Table 9.1. Common stable isotopes

Hydrogen	^1H, ^2H
Carbon	^{12}C, ^{13}C
Nitrogen	^{14}N, ^{15}N
Oxygen	^{16}O, ^{17}O, ^{18}O
Sulfur	^{32}S, ^{33}S, ^{34}S, ^{36}S

In general, chemistry is driven by the number of electrons a substance has. However, the weight of a substance can make a difference as well. For instance, if a chemical reaction of hydrogen involves the frequent exchange of hydrogen between two media (say, a solid and a gas), the difference in weight between the two isotopes of hydrogen (^1H and ^2H) can be significant as ^2H is effectively double the weight of ^1H. If weight becomes an issue, the heavier isotope (^2H) will eventually participate in the chemical reactions less, and the resultant mixture will become fractionated with respect to its isotopes.

In general (though not always): a chemical system that consists of two reservoirs—an active one and an inactive one—will become fractionated with the lighter isotope becoming concentrated in the active reservoir and the inactive reservoir holding more of the heavier isotope. This is called mass-dependent fractionation. Mass-dependent fractionation can be extremely sensitive to temperature. In general, fractionation increases with decreasing temperature. As a result, isotopes can be used as thermometers, and often detail the formation temperatures of certain systems.

9.2 Fractionation

The isotopic fractionation of a system can be represented a few ways. One way is the fractionation factor, α. This value is equal to:

$$\alpha = \frac{\left(\frac{^BX}{^AX}\right)_1}{\left(\frac{^BX}{^AX}\right)_2} \quad (9.1)$$

Where ^BX is isotope of element X with weight B, ^AX is the isotope with weight A (usually lighter than B), and 1 and 2 denote two different reservoirs. For instance, for at 20°C, the ratio of ^{18}O and ^{16}O in water between vapor and gas is:

$$\alpha = \frac{\left(\frac{^{18}O}{^{16}O}\right)_{Vapor}}{\left(\frac{^{18}O}{^{16}O}\right)_{Liquid}} = 0.99$$

Since α is less than 1, this implies that the liquid is more enriched in the heavier isotope than the vapor.

More frequently used than the α value or fractionation factor is the delta (δ) value. This has as its general formula:

$$\delta = 1000 \times \frac{\left(\frac{^BX}{^AX}\right)_{sample} - \left(\frac{^BX}{^AX}\right)_{standard}}{\left(\frac{^BX}{^AX}\right)_{standard}} \quad (9.2)$$

The δ value uses standard isotope ratios for its calculation. These standards vary by isotope and are agreed upon by the isotopic geochemistry community. Standards are listed as table 9.2. A negative δ value implies that a sample is depleted in the heavier isotope (B) than the standard, whereas a positive value implies the opposite.

Table 9.2. Isotopic standards.

Hydrogen	Standard Mean Ocean Water (SMOW)
Carbon	Pee-Dee Belemnite
Nitrogen	Air
Oxygen	Standard Mean Ocean Water (SMOW)
Sulfur	Canyon Diablo Troilite (FeS)

9.3 Identification of origin.

One of the principle uses of stable isotope geochemistry is in the identification of origin of a material. Some processes are better at fractionating elements than others. Life is a process that readily fractionations carbon. Organic material from life is usually depleted in ^{13}C.

As an example, some of the oldest metasedimentary rocks are found in Greenland, and are known as the Isua greenstone belt. Within these rocks were reported grains of graphite that showed an isotopic depletion in ^{13}C. They had a $\delta^{13}C$ value of about -30, though what exactly this records is unclear (for more detail, refer to Mojzsis et al. 1996 Nature 384, 55-59).

Another application of using isotopes for identification of origin is in the use of nitrogen and oxygen isotopes to identify sources of nitrate in water. The source of nitrate can be determined based on its isotopic composition (table 9.3). This tool allows researchers to determine what sources are contaminating an aquifer. For more detail see http://wwwrcamnl.wr.usgs.gov/isoig/guidelines/nitrate/Overview.htm

Table 9.3. Nitrate isotopes and origins.

$\delta^{15}N$	$\delta^{18}O$	Source
-4 to +4	+23	Fertilizer made from air
+15±5	-5 to +15	Animal waste
+4 to +9	+1 to -4	Soil Nitrogen
Up to +40	Up to +20	Nitrate from denitrification
-3	+18 to +60	Precipitation

Isotopes feature prominently in cosmochemistry. Organic material in meteorites often bears strong enrichments in the heavy isotopes of C, N, and H. These strong enrichments are indicative of the temperature and location where these materials formed. Isotopes are also used to identify meteorites from other planets. This is done by using the three isotopes of oxygen, ^{16}O, ^{17}O and ^{18}O. Nearly all terrestrial processes that result in fractionation affect the $\delta^{18}O$ value twice as much as the $\delta^{17}O$ value, because the mass difference between ^{18}O and ^{16}O is twice that of the ^{17}O to ^{16}O difference. When the oxygen isotopes of meteorites are examined, very few fall on the terrestrial fractionation line (where nearly all earth rocks fall). As a result, oxygen isotopes are key to identifying an extraterrestrial origin for several meteorites (Figure 9.1).

Figure 9.1. Oxygen isotopes for meteorites. From http://solarsystem.nasa.gov/scitech/display.cfm?ST_ID=2398

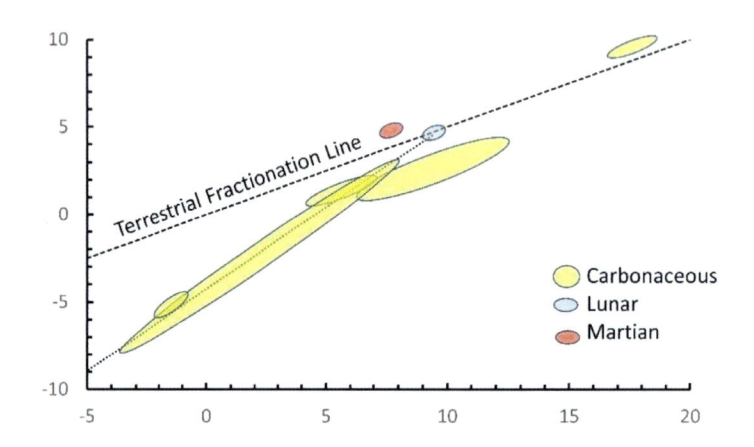

9.4 Paleoecology

A major use of isotopes is determining the ecology of organisms (e.g., what eats what?). In general, the higher up the food chain you go, the heavier the isotopic composition. This is because proteins and body material are enriched in heavier isotopes as these materials are the "less active" reservoir in biochemistry. As an organism processes chemical energy, lighter isotopes are burned off and removed as waste or respiration products, whereas the heavier isotopes remain in the flesh. If an organism eats another creature, then the organism now has as its food source an isotopically heavier material. As it respires and excretes as waste some of these compounds, its flesh becomes even more enriched in heavy material. This can proceed with each step of a food chain.

In addition to herbivore/carnivore determination, isotopes can be used to determine paleoecologic climates. This uses differences in isotope compositions of plants to arrive at a common ecosystem. For instance, there are two types of photosynthetic pathways plants use to turn CO_2 into sugars: C3 and C4 (among others). The C3 pathway is used at lower temperatures under cool, moist conditions. It results in significant fractionation of ^{13}C from ^{12}C, and has a $\delta^{13}C$ value of about -25. The C4 pathway is used at higher temperatures in dryer environments, and results in less fractionation with a $\delta^{13}C$ value of about -10. Using this information, the ecology of an environment can be reconstructed from herbivores.

9.5 Water Cycle and Paleothermometry

Isotopes are especially useful in the water cycle, both to identify latitude for material, and to identify temperatures (these two are obviously linked). As you go closer to the poles of the globe, the fractionation of ^2H from ^1H and ^{18}O from ^{16}O both increases because of cooling of the environment. As a result, precipitation in these locations is consistently depleted in the heavy isotopes and shows negative δ values that become more negative as you approach the poles (figure 9.2).

Figure 9.2. Schematic of ^{18}O delta values of precipitation with respect to latitude

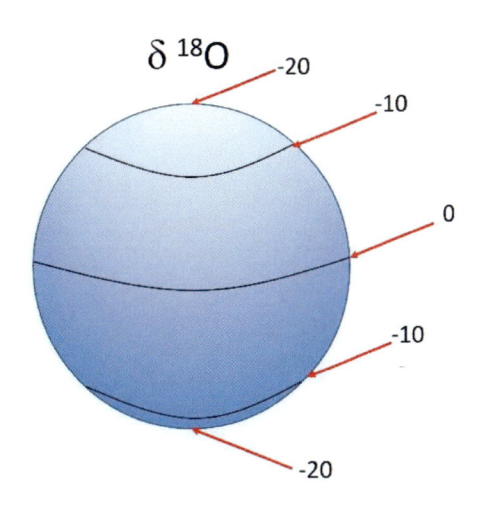

The isotope composition of surface water is also useful for elucidating temperature. As precipitation become lighter in isotopic composition (as temperature cools), the water that remains at the surface becomes heavier. As a result, material from this water (for instance, ice or forams) becomes enriched in the heavier isotopes.

Chapter 9 Practice Problems

1. A broken stalactite has an $^{18}O/^{16}O$ ratio of 0.002009. SMOW has a ratio of 0.002005. Calculate the delta value.

2. A nitrate has a $\delta^{18}O$ value of +5 and a $\delta^{15}N$ value of +15. What is its likely source?

3. An organism has a $\delta^{13}C$ of -8 and a $\delta^{15}N$ of +12. What is its diet? Use the following chart:

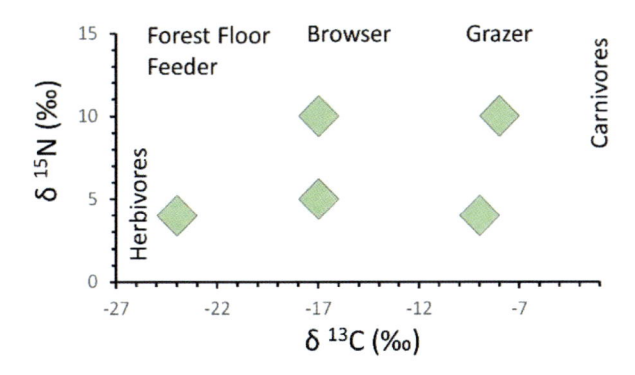

4. A carbonate has a $\delta^{18}O$ of +1. What was the temperature at which it formed (modified from Epstein et al. 1953)?

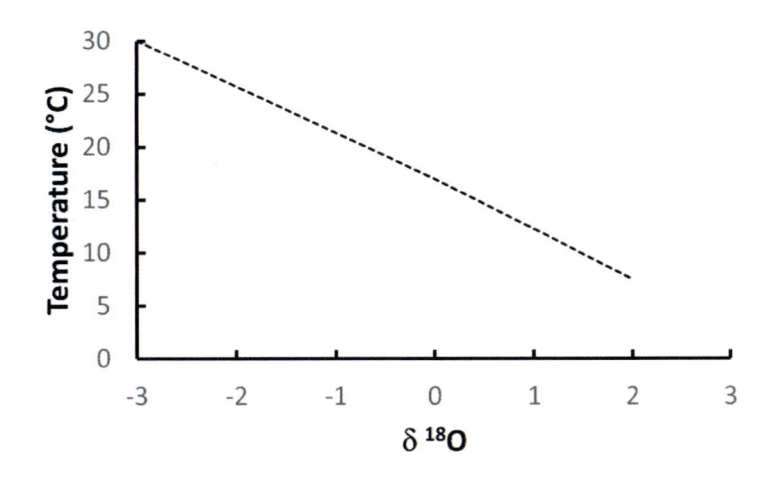

5. Why does meteoritic material bear such strong enrichments in heavy isotopes of H, C, and N?

Chapter 10. Water-Rock Interactions 1. Metasomatism

10.1 Water-rock interactions

Water is an important substance in geology. Water can modify rocks in a substantial fashion. The results of water-rock interactions include the formation of important ores and mineable substances, the shaping of geomorphic features, and the flow of water through the subsurface.

This is largely due to water's excellent solutional properties. Many ions readily dissolve in water, including rock forming ions such as Na^+, K^+, and Ca^{2+}. As a result, when rocks bearing these elements are in contact with water, these elements slowly dissolve into the water, stripping them from the rock and changing the composition of the rock (white circles, Figure 10.1). Additionally, the water may have dissolved constituents itself, resulting in deposition (green circles, Figure 10.1) or replacement (blue circles, Figure 10.1) of some minerals as it flows through the rock. Other times the water may itself become incorporated into the rock, generating new minerals with new properties.

Figure 10.1. Water rock interactions schematic

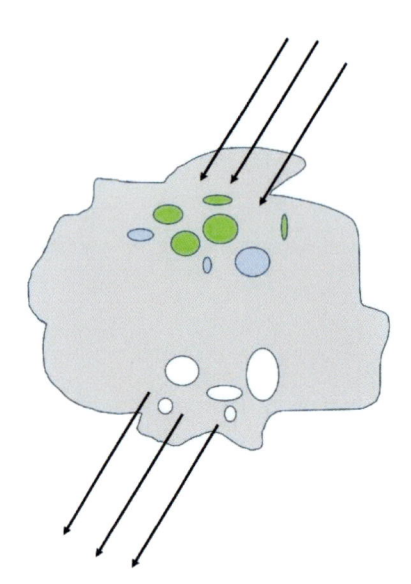

Water-rock interactions are an important part of the geochemistry of the earth's surface. The next few chapters will discuss four different types of water-rock interactions: metasomatism, weathering, karst, and diagenesis.

10.2 Water in metamorphism

Water is an important constituent in metamorphic processes. The action of water is important to determining the mineralogy present, such as the formation of clays, phyllosilicates, and amphiboles. However, in a classical sense, metamorphism is a closed system- that is, the major elements present remain present during metamorphism. In classical metamorphism, mineralogy may change, but the bulk composition does not, except for volatile molecules such as CO_2 and H_2O.

Metasomatism is a variety of metamorphism that allows for the changing of bulk composition, usually by the action of hot water. As hot water flows through a rock, it strips out some elements from the rock, and deposits these elements elsewhere as the water properties change (for instance, as it cools). This hot water also reacts with the rock, generating new minerals, several of which are of economic importance. Indeed, much of the earth's major noble metal deposits (gold and platinum) are linked to metasomatic processes occurring to the rocks at the earth's surface.

10.3 Serpentinization

An important variety of water-rock interaction on earth, and likely elsewhere in the universe, is the process of serpentinization. In general, serpentinization comes about as mafic rocks (Mg- and Fe-rich igneous rocks, typically composed of olivine and pyroxene) react with hot water to generate new minerals and change in redox state. Olivine is the major player in the serpentinization reaction.

Recall that there are two compositional end-members of olivine, forsterite Mg_2SiO_4, and fayalite Fe_2SiO_4. Any given occurrence of olivine typically has some mixture of these two components; because as minerals they are completely interchangeable in crystal structure. These two end members react differently with water, however:

$$3\ Fe_2SiO_4 + 2\ H_2O = 2\ Fe_3O_4 + 3\ SiO_2 + 2\ H_2$$

$$3\ Mg_2SiO_4 + 4\ H_2O + SiO_2 = 2\ Mg_3Si_2O_5(OH)_4$$

Fayalite reacts with water to generate magnetite and free silica, in addition to H_2 gas. Forsterite reacts with the free silica and water to generate a serpentine mineral. Serpentine minerals include asbestiform minerals; serpentinites are the main sources of these industrially important, much-maligned minerals. Additionally, the serpentine mineral itself can react to with carbon dioxide in the environment:

$$2\ Mg_3Si_2O_5(OH)_4 + 3\ CO_2 = Mg_3Si_4O_{10}(OH)_2 + 3\ MgCO_3 + 3\ H_2O$$

The results of this reaction are a magnesium carbonate, and talc.

These chemical reactions (and there are others) result in several changes to the geochemical environment:

1) In most cases, serpentinization is an exothermic process. Some of the deep sea hot vents in the ocean are driven by serpentinization, and the process ends up heating the water.
2) Additionally, olivine, with a density of 3.3 g/cm^3 changes to serpentine minerals, with densities of 2.7 g/cm^3. As a result, there is a volumetric expansion of the rock, which destroys primary structures and can cause fracturing in the surrounding rock.
3) When serpentinized, the more ferroan olivine releases significant quantities of H_2, a strong reducing agent. Reduction of CO_2 by H_2 can form methane CH_4, and metals such as nickel and even native iron can form because of this H_2. Reduction of CO_2 to methane may be the source of Martian methane. Formation of iron and nickel metal in serpentinization these rocks can also trap platinum group elements, and serpentinites are one source of these valuable metals.
4) In many cases, serpentinization is accompanied by formation of $MgCO_3$ or $Mg(OH)_2$. These minerals increase the local pH to up to 10-11, resulting in strongly alkaline water.

10.4 Skarns

As molten rock moves through crustal carbonate rock, it heats the rock, liberating water from the crustal rock (Figure 10.2, left panel). This hot water (as steam or a superheated fluid) reacts with the crustal rock and igneous material, allowing some elements to migrate into the fluid. The elements that move into the fluid are those elements that are highly soluble in this water, or that are incompatible in the carbonate minerals that mostly comprise the crustal rock. These incompatible or soluble elements eventually settle around the igneous body as it cools, forming a rind known as a skarn (green layer in right panel, Figure 10.2). Skarns can be ore bodies, as incompatible elements include copper, silver and gold.

Figure 10.2. Skarn formation schematic.

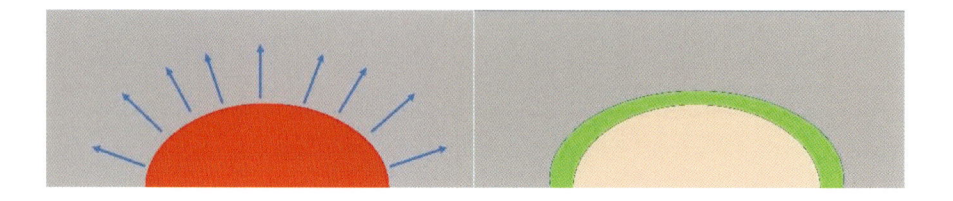

10.5 Hydrothermal leaching of rocks

A typical crustal rock consists of Ca, Al, Si, Na, K, Mg, Fe, and O as major elements. If a typical crustal rock reacts with water, some of the elements move into the water. The first elements to be lost include Na and K, followed by Ca and Mg. Depending on the conditions of the water, including pH and sulfur content, other elements can be stripped, including Fe and even Al.

Hydrothermal vents are sometimes associated with volcanic island arcs. These hydrothermal vents are characterized by low pH, high sulfate waters. These waters can strip elements from rocks, leaving a mixture of Al and Si oxides and hydrates. Many of these high-sulfur (referring to the high oxidation state of sulfur) hydrothermal vents are surrounded by silica and aluminous clays such as kaolinite $Al_2Si_2O_5(OH)_4$. These minerals in turn can be metamorphosed to form aluminosilicate minerals such as kyanite, an industrially important mineral (Figure 10.3).

Figure 10.3 Kyanite quartzite from Willis Mountain, Virginia, USA. They kyanite (white blades) is embedded in quartz (gray mineral). This rock is mined to produce spark plugs. For more information, see Owens and Pasek (*Econ. Geol.* 2007).

Chapter 10 Practice Problems

1. If 1.00 kg of olivine (assume 75% forsterite, remainder fayalite) reacts with water how much H_2 would you expect to form?

2. Fayalite can react with dissolved CO_2 to form methane CH_4, magnetite and silica. Balance this reaction.

3. Brucite $Mg(OH)_2$ is a common mineral associated with serpentinization. What pH would you expect for water in direct contact with brucite? Assume the temperature of the water is 298 K and the concentration of Mg^{2+} in the water is 0.05 M (ocean water concentration).

Use the thermodynamic data here:

	ΔG (kcal)
$Mg(OH)_2$	-199.39
Mg^{2+}	-108.7
H^+	0
H_2O	-56.687
OH^-	-37.594

Chapter 11. Water-Rock Interactions 2. Weathering

11.1 Water-rock interactions and weathering

As the prior chapter showed, water is an important substance in geology, especially in the formation and transport of economically valuable materials. In addition to transporting key elements at high temperature (100-300 °C), water is important in altering and changing rocks at a much lower temperature as well (at 25 °C, and less). At these lower temperatures, water is still important in changing the bulk composition of rocks, again due to water's excellent solution properties. Many ions readily dissolve in water, including rock forming ions such as Na^+, K^+, and Ca^{2+}. As a result, when rocks bearing these elements are in contact with water, these elements slowly dissolve into the water, stripping them from the rock and changing the composition of the rock. Additionally, the water may have dissolved constituents itself, resulting in deposition or replacement of some minerals as it flows through the rock. Other times the water may itself become incorporated into the rock, generating new minerals with new properties.

Water is especially important in the changing of the surface of the earth through weathering. By weathering rocks, mountains that were once among the tallest on the earth slowly shrink to hills, volcanoes turn into fertile soil, and important ore bodies for some major elements are formed.

11.2 Types of Weathering

Weathering can be broadly divided into three varieties: chemical, physical, and biological. Very few types of weathering are solely of one variety, however. Most physical and biological weathering involves some variety of chemistry (for instance, phases changes of water) and even chemical weathering involves a fair bit of physical abrasion and is often reliant on biology to proceed.

Despite this, historically weathering has been classified into the two or three varieties mentioned above. Physical weathering includes processes linked to temperature changes, abrasion by materials, and expansion and contraction of material within cracks. Types of physical weathering include *Aeolian* weathering, which is weathering caused by abrasion of rocks by wind-blown sand and dust. If you've been to the American southwest, you've probably seen features formed by Aeolian weathering. *Pressure release* weathering occurs when a rock formation that was previously burdened by an overlying layer of material no longer has that material and hence rebounds from lack of stress. *Thermal expansion* occurs to rocks that are heated (and thereby expand) causing fracturing and weakening of the rock. Heating can be due to forest fires, or even day-night cycles. *Frost weathering* occurs when water within cracks of a rock freezes, expanding the cracks. There are a few varieties of weathering linked to frost, depending on the climate of rock weathering. *Salt expansion* occurs as salt-filled water dries and results in the crystallization of salt, which causes pressure at the grain boundaries and fractures the rock.

Biological weathering includes several processes linked to action by organisms. Many involve substantial biochemistry, although a few are solely physical in nature. For instance, if mountain goats tread a single path for multiple generations, the path wears away because of being hit by their hooves. Other biological weathering processes include breakdown of rocks by lichen and tree roots, which as they grow force the expansion of rock cracks and promote rock dissolution by releasing organic acids. One variety of biological weathering that is especially noticeable (and observed by the author on the beaches of California) is the burrowing of clams into rocks (Figure 11.1). This process is primarily mechanical in origin but also includes some chemical changes to the rocks as the clams burrow.

Figure 11.1. Clam burrows from Ventura Beach, California

11.3 Types of Chemical Weathering

As this book and class focus primarily on chemistry, we shall also focus primarily on chemical weathering. Most chemical weathering is promoted by water, as is most physical weathering. Again, water is an important part of weathering.

Hydration is a chemical process by which water becomes part of the chemical structure of a rock. Hydrated minerals within the resulting rock are generally less dense, and as a result, put pressure on a rock and cause it to fracture (e.g., Figure 11.2).

Figure 11.2. As part of a rock becomes hydrated, the volumetric expansion results in cracking of the initial rock.

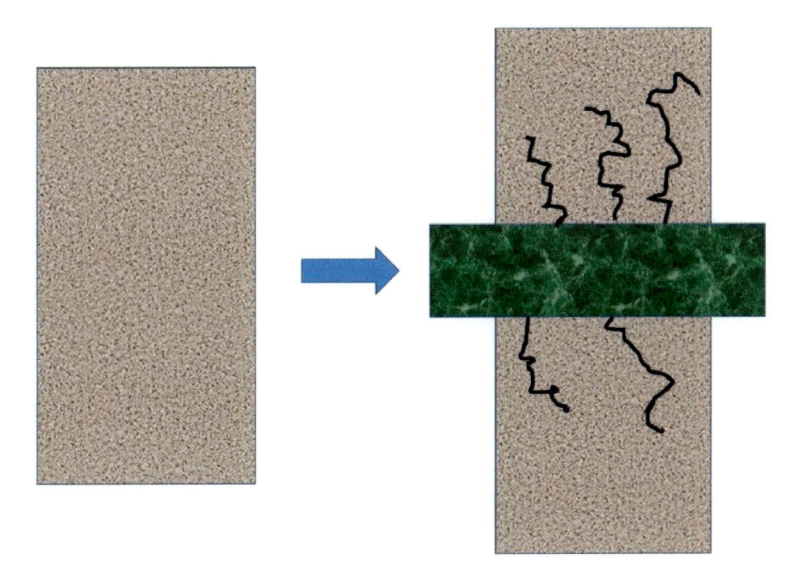

Hydrolysis (not to be confused with hydration!) is the process by which water becomes part of the chemical structure of a rock, and then by which the water removes some of the more soluble components of the rock. Typically, the resulting minerals of this process are weaker than the beginning minerals, and hence the rock is much more susceptible to weathering by physical processes. Hydrolysis and hydration are closely related.

Dissolution is the process by which water reacts with a rock to dissolve away parts of the rock. This is an important process the economic geology of aluminum. Aluminum is one of the least soluble major elements that make up most rocks, and as a result, in highly weathered rocks, aluminous deposits known as bauxite are one of the world's primary sources of aluminum ore.

Carbonation is the process by which atmospheric CO_2 reacts with water to form carbonate and acid. This acid promotes the dissolution of surficial rock and is very important in karst geochemistry, which we will cover next.

Oxidation is the process by which a reduced mineral becomes oxidized and typically expands, resulting in fracturing of the rock. We've already covered oxidation extensively, but if you consider the oxidation of iron in magnetite to hematite:

$$2\ Fe_3O_4 + 0.5\ O_2 = 3\ Fe_2O_3$$

You can see that the minerals went from having fewer oxygens to having more oxygens as oxidation occurs. As a result, the mineral expands (which makes intuitive sense as the final mineral has more atoms in its unit structure from reaction with O_2 as a gas).

Chemical biological weathering is the process by which organisms weather rocks, either by the addition of acid, promotion of oxidation, or chelation of cations (bonding of cations such as Ca^{2+} and Mg^{2+} with acids such as acetate). This process includes many of the above mentioned chemical weathering processes, but have been specifically identified to be promoted by biology, and hence are separated by the "biological" term. It has been proposed, reasonably, that most chemical weathering processes are promoted by biology, specifically the action of microbes. For *oxidation* this seems reasonable, but other varieties may not be always true.

Chapter 11 Practice Problems

1. If an average granitic rock is 15 weight percent Al_2O_3, how much rock must weather away to form a bauxite deposit?

2. Determine the pH (4 or 8) at which the following hydrolysis reaction occurs best: $3NaAlSi_3O_8 + 12H_2O + H^+ \rightarrow Na_2Al_3Si_3O_{10}$ x $2H_2O + Na^+ + 6H_4SiO_4$. Assume the Na^+ concentration is 0.5 M and dissolved silica is 1 ☐M.

3. If magnetite weathers to hematite through an oxidation reaction mediated by water, what volume change would occur to a 1 cm^3 crystal of magnetite as it oxidizes? The density of magnetite is 5.15 g/cm^3 and the density of hematite is 5.27 g/cm^3

Chapter 12. Water-Rock Interactions 3. Karst

12.1 Karst Geology

A general theme of these last two chapters has been how water is an important substance in geology, especially in the formation and transport of economically valuable materials. In addition to transporting key elements at high temperature (100-300 °C), water is important in altering and changing rocks at a much lower temperature as well (at 25 °C, and less). At these lower temperatures, water is still important in changing the bulk composition of rocks, again due to water's excellent solution properties. This chapter covers the specific case of water interacting with the surface and subsurface to dissolve and precipitate rocks, focusing primarily on the dissolution of limestone, which is dominated by calcium carbonate minerals ($CaCO_3$) such as calcite and aragonite, as well as dolomite ($CaMg(CO_3)_2$). Calcium carbonate readily dissolves in undersaturated water, and precipitates out of supersaturated water.

The morphological landscape defined by the dissolution and precipitation of near-surface bedrock is called Karst. This morphology primarily occurs in carbonates (but not always) and is defined by specific dissolution features throughout the surface and subsurface, including caves and sinkholes, as well as water conduits such as sinks, swallets, and springs.

12.2. Karst Geochemistry

Karst in limestone occurs as limestone dissolves away through the action of water. Four chemical reactions define the dissolution of karst:

$$CaCO_3 \rightarrow Ca^{2+} + CO_3^{2-} \qquad K = 5 \times 10^{-10} \qquad (12.1)$$

$$CaCO_3 + H^+ \rightarrow Ca^{2+} + HCO_3^- \qquad K = 10^2 \qquad (12.2)$$

$$CaCO_3 + 2H^+ \rightarrow Ca^{2+} + H_2O + CO_2 \qquad K = 3 \times 10^8 \qquad (12.3)$$

$$CaCO_3 + H_2O + CO_2 \rightarrow Ca^{2+} + 2HCO_3^- \qquad K = 3.4 \times 10^{-5} \qquad (12.4)$$

In general, reaction (12.1) is more important at high pH (basic conditions), and reaction (12.3) is more important at lower pH (acidic conditions), with reaction (12.2) being important in between these two. Reaction (12.4) is the most important for water in direct contact with the atmosphere, and drives karst geology significantly. The CO_2 in reaction (12.4) comes directly from the atmosphere by dissolution of atmospheric CO_2.

Water is considered to be "corrosive" if it is capable of dissolving carbonates. Several factors go into determining whether water is corrosive, including carbon dioxide concentration, temperature, pH, calcium and carbonate content, and organic material and contaminant content.

In general, water that has significant amounts of dissolved CO_2 is more corrosive than water without, resulting from action of reaction (12.4). Most surface water, being in direct contact with the atmosphere (and about 400 ppmv of CO_2) is more corrosive than subsurface water. As a result, karst landforms that occur at the surface expand more quickly than those that occur at the subsurface.

Calcite is more soluble at lower temperatures than it is at higher temperatures (Figure 12.1). As a result, cooler climes (and seasons) promote limestone dissolution.

Figure 12.1. Temperature dependence of calcite dissolution. The Y axis shows the K of the reaction, and the X axis is temperature in Celsius.

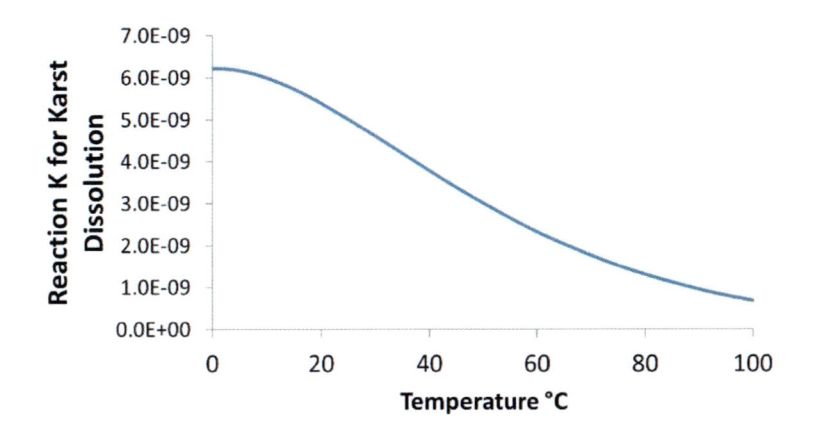

The effect of pH on limestone dissolution is obvious from reactions 12.2 and 12.3, and is the subject of the practice questions below.

Water can only bear so much calcium and carbonate before it is saturated with these two components. Water that has no calcium or carbonate is hence quite corrosive, and precipitation or rain, bearing no significant calcium, readily promotes dissolution of carbonates in karst.

Contaminants including industrial and organic acids, some of which can *chelate* calcium. *Chelation* is the process by which an organic compound binds to a cation in solution, removing it from reaction. Acetate and EDTA are common chelating agents, and promote the dissolution of calcite by removing $[Ca^{2+}]$ from solution.

12.3 Karst Landforms

Karst is characterized by several unique landforms, most famously caves, but also sinkholes (and the potential for loss of life) and springs.

Springs. A spring is a location where groundwater naturally erupts to the surface. This is contrasted to an *Artesian well*, a location where groundwater erupts to the surface as the result of drilling a well. A spring can form as the result of several processes, including faults, which provide a crack for water to flow through, and cave systems, which provide a conduit for groundwater (Figure 12.2). Some springs, known as *seep springs* are the result of diffuse flow at the surface and do not have a single tunnel or point where the water erupts out from the ground.

Figure 12.2 Ginnie Springs, located in northern peninsular Florida, forms as a cave-conduit surfaces from the ground.

Swallets, Windows, and Rises. Within the karst environment, sometimes water will enter the ground at a specific opening known as a swallet. A swallet may in turn feed into a conduit, and the conduit may erupt at the surface at a point downstream, where it results in a rise. A rise differs from a spring in that it has a known point of entry for the water, and has not seen its water filtered through bedrock. Sometimes underground rivers and conduits may be visible from the surface at a certain point, but do not flow in or of out of the surface at that point, resulting in a karst window.

Caves are formed in karst environments through the dissolution of bedrock by corrosive water (Figure 12.3, blue arrows). The bedrock dissolves away, leaving a void. Caves are mostly found in limestone or dolostone sedimentary environments where they form because of the chemical reactions described above, though caves can also form in ice, in lava (as lava tubes), and even in sandstone, though they are less common in those environments. Caves can be full of water, or may be dry, though an active cave (one undergoing precipitation and dissolution of carbonates) will be wet. *Speleothems* are cave formations that include stalactites growing from the cave ceiling, and stalagmites growing from the cave floor, among many others (too many to list here, and varying by formation process and morphology, see some examples as Figure 12.4). Many of these form as the result of water saturated with carbonate and calcium entering the void space of a cave and evaporating slightly. The evaporation leads to supersaturation of these ions, and precipitate out calcite. This process may be biologically-mediated, hence these cave formations may be similar to stromatolites or other microbial biofilms.

Figure 12.3 Schematic of cave formation.

Figure 12.4. Speleothems in Luray Caverns, Virginia.

Sinkholes can be broadly considered to be the intersection of a collapsed cave ceiling with the surface of the earth. They are responsible for much small-scale destruction in karst environments. Sinkholes form primarily through the weakening of a cave ceiling, which is enhanced by pumping groundwater (removing water to support the cave), vibrations (weakening the rocks), flow of corrosive water into the cave (making it bigger), and changes in water level.

Chapter 12 Practice Problems

1) What is the pH dependence of reaction 2 and 3? Which reaction is more important at a pH of 3? Of 6? Of 9?

2) Global warming is due in part to rising CO_2 concentrations. In 2013, the amount of CO_2 in the atmosphere reached 400 ppmv (parts per million by volume). In contrast to the pre-industrial age CO_2 of ~200 ppmv, how much more soluble is calcite in water (in moles/L) in direct contact with the atmosphere?

3) If a water sample from a cave has 10^{-5} M of Ca^{2+} and 2×10^{-5} M of CO_3^{2-}, what percent of the water would need to evaporate to form $CaCO_3$?

Chapter 13. Water-rock interactions 4. Diagenesis

13.1 Diagenesis

Our final chapter in the water-rock interactions section is the low-temperature change of geologic material through time, usually mediated by water. This process is broadly known as *diagenesis*, and can be summarized as "how sediments become rock". The transformation of sediments to rock occurs at lower temperature (100-300 °C) with water playing an important role in the transport of elements. Again, this is because of water's excellent solution properties. There is a gradient between diagenesis and metamorphism, and the boundary between the two is not necessarily clear, but diagenesis occurs at lower temperatures and pressures. This process is a fundamental part of sedimentology and is often covered there as part of that course.

13.2 Diagenetic processes

There are many processes that make up diagenesis, including compaction, bioturbation, replacement, dissolution, and precipitation. In general, if you recognize the word from earlier in this book, the meaning is still the same. Note that this is not an exhaustive list, and there are other processes that are diagenetic, including petrification, dolomitization, silicification, and redox changes.

If you compare a sedimentary rock with its sediment precursor, you will notice several (obvious) things. For one, a rock is harder and stronger. This is because sedimentary grains have been glued together, often by cementation. Cementation is a process driven by dissolution of grains at the boundaries in part, and precipitation of a cement such as a carbonate or silica. Additionally, the sediments have been compacted, and have lost water. The dehydration of a sediment is important in formation of a sedimentary rock. Additionally, a sediment may be reworked prior to undergoing cementation. Burrowing animals may alter and destroy layering, resulting in changes to the sedimentary structure of the resulting rock. Note that these processes occur early in diagenesis.

After the sediments have become a solid, further changes may take place. These diagenetic changes include dissolution and precipitation of new minerals (for instance, dissolving a calcium carbonate shell away within a rock, and filling the void with silica), reduction of porosity by precipitation of minerals at grain boundaries, and changes to clay and carbonate minerals (for instance, transformation of smectite to illite). These processes occur after the rock has already formed generally, as they are slower than the primary genetic processes, and take place as the rock is part of the geologic column.

Once the rock is uplifted and exposed to weathering, further changes may take place, though many of these overlap with weathering. These changes include oxidation of rock minerals (such as transformation of pyrite into iron oxides, sulfate and acid), dissolution of carbonates and other soluble minerals, and transformations of other minerals such as clays.

13.3 Types of Chemical Reactions

Chemical reactions that take place in diagenesis are identical to those that we have discussed previously. <u>Dissolution</u> is the process by which a solid enters the liquid phase, for instance the dissolution of silica:

$$SiO_2 + 2H_2O \rightarrow H_4SiO_4 \qquad \text{or} \qquad SiO_2(s) \rightarrow SiO_2(aq)$$

Water samples that have less than the equilibrium amount of dissolved SiO_2 (or silicic acid) are called *undersaturated* with respect to silica, and hence will dissolve silica when in contact with this mineral. Mineral solubility is governed by several factors, including temperature, pH, the ionic strength of the water, and the pressure of the fluid.

Water that has too much of a chemical constituent will result in <u>precipitation</u>, the process by which a solute leaves the liquid phase to become a solid. As an example, the precipitation of gypsum:

$$Ca^{2+} + SO_4^{2-} + 2H_2O \rightarrow CaSO_4 \times 2H_2O$$

occurs when a water has more Ca^{2+} and SO_4^{2-} than is tolerated at equilibrium. Such a solution is called *oversaturated* with respect to gypsum.

<u>Mineral conversions</u> are changes to minerals that take place as a result of thermodynamic instability. The most notable change to occur is the transformation of aragonite to calcite. Both are minerals with the formula $CaCO_3$, but there is a difference in the thermodynamic stability between these two minerals, and thus, if one of these minerals is present in a fossil, it will often change to the other mineral during diagenesis (which mineral will turn into which is a question below).

<u>Exchange reactions</u> are reactions that trade one cation or anion in a mineral for another in the reacting solution. The best example of this process is the formation of dolomite:

$$2CaCO_3 + Mg^{2+} \rightarrow CaMg(CO_3)_2 + Ca^{2+}$$

This reaction occurs commonly where calcium carbonate interacts with Mg-rich water, such as sea water.

Redox reactions are those that we have covered extensively in prior chapters. Examples of diagenetic redox reactions include the oxidation of pyrite within sedimentary rocks. Pyrite may form within sedimentary rocks as the result of precipitation from a reduced fluid. This pyrite may subsequently oxidize, resulting in a change of minerals and often of volume as well.

Chapter 13 Practice Problems

1) A geode is formed by the precipitation of SiO_2 from water in a hollow, spherical void within a rock. If you find a geode with a radius of 4 cm, is the water within the void able to form a geode with 1 cm long quartz crystals solely by evaporating from a saturated solution? Assume the solubility of SiO_2 is 100 mg/L at saturation (and the temperature of interest) and the density of SiO_2 is 2.65 g/cm³. What volume of water is necessary to form a geode with crystals 10 cm long in a 1 m radius geode?

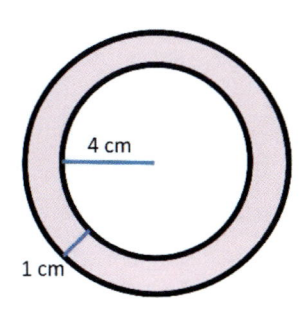

2) Of aragonite and calcite, which is the more stable? How do you know?

You can use these thermodynamic data:

Mineral	ΔG^0 (kcal/mole)
$CaCO_3$ (aragonite)	-269.55
$CaCO_3$ (calcite)	-269.80

Chapter 14. Properties of Groundwater

14.1 Groundwater

One of the most important parts of aqueous geochemistry (and one of the easiest ways to get paid as a geologist) is groundwater geochemistry. Groundwater is water that is extracted from the subsurface. It is important to many geographical regions, as surface water accumulation is often seasonal whereas water needs are not.

A water-bearing subsurface layer within the earth is called an aquifer, whereas one that is poorly permeable to water is called an aquiclude. Many areas have a surficial aquifer, one that is not bounded on the top by an aquiclude. Precipitation directly feeds into this aquifer.

Groundwater geochemistry is its own area of study, like many of the topics we have covered. However, the chemistry of groundwater is characterized by terms that should already be familiar to us as geochemists.

14.2 Properties of Groundwater

If you take a sample of groundwater there are several features of that water you will measure. Four features measured for almost all groundwater samples (and surface water samples) are temperature, pH, ORP, and TDS. These are often complemented by an in-depth examination of hardness and composition.

The **pH** of a groundwater is a fundamental property of groundwater, and is, as we have stated before, the $-\log_{10}$ of the H^+ concentration. Since pH is somewhat dependent on temperature, a more useful view of pH is the excess of H^+ with respect to OH^-.

The **Temperature** of water determines mineral solubility, gas solubility, and general reactivity. Any of the three temperature scales (°F, °C, K) may be used for this purpose, though °F will only be in the US, and K will usually only be used by academic institutions.

The **Oxidation-Reduction Potential** (ORP) of a water determines the Eh of that fluid, making Eh-pH diagrams useful. In general, ORP is reported in mV, and it should correlate to dissolved oxygen content (sometimes abbreviated as DOC, though dissolved organic content is often also termed DOC). The dissolved oxygen content is generally less useful than ORP, as ORP is easier to measure.

The **Total Dissolved Solids** (TDS) of a groundwater sample determines its potability and salinity. Most of the time, TDS is recorded as **Electrical Conductivity (EC).** EC, which has units of µS/cm, is somewhat obscure to chemists, but comes from how conductive a water is. The more conductive it is, the saltier it is. In general, the EC (µS/cm) × 2/3 = TDS (ppm).

The **Composition** of a groundwater is determined by the material dissolved within it. In most cases, these are the cations Mg^{2+}, Na^+, K^+, and Ca^{2+}, as well as the anions Cl^-, SO_4^{2-}, HCO_3^- and CO_3^{2-}. These ions determine water type.

14.3 What determines the composition of groundwater?

The composition of groundwater is partially determined by the composition of the aquifer. That is, if the aquifer is made of limestone, then the composition will likely be rich in Ca^{2+} and HCO_3^-/CO_3^{2-}. If the aquifer is in contact with evaporites, you may expect to see sulfate and Mg^{2+} instead.

In addition, groundwater composition is determined by temperature, which affects solubility of minerals and gases, as well as age (older water will have equilibrated with the surrounding aquifer more so than younger water). Additionally, the composition of a groundwater sample can be strongly influenced by the amount of salt water in contact with the aquifer. Finally, pollution, anthropogenic effects, and source material can all affect groundwater composition.

14.4 Piper Diagram

The Piper diagram is a construct used in groundwater geochemistry to understand the origin of a groundwater and the relationship between different groundwater samples. The piper diagram is a diagram that consists of two ternary (triangle) diagrams that then translate onto a parallelogram (Figure 14.1).

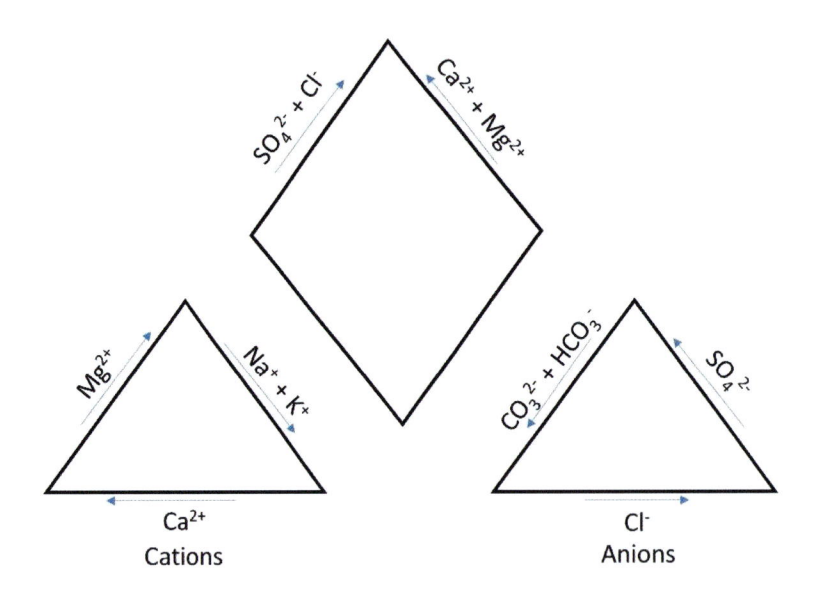

To make a Piper diagram, you must do the following:

1) Convert all your units to meq/L (milliequivalents per liter). You do this by converting first to mM (millimolar), and then multiplying each species by the absolute value of the charge (e.g., 1 for Na^+, 2 for SO_4^{2-})

2) Add up all the values for cations, and then determine the relative proportion of each. For instance, if Na^+ and K^+ are together 1 and Ca^{2+} is 8, and Mg^{2+} is 1 (in meq/L), then Na + K is 10%, Ca is 80%, and Mg is 10% of the total charge.

3) Plot these on the ternary diagram for cations (practice this in mineralogy or petrology).

4) Do this again for anions.

5) Determine the intersection of the points from cations and anions, and plot the point on the parallelogram.

To make your own Piper diagrams, go to: http://water.usgs.gov/nrp/gwsoftware/GW_Chart/GW_Chart.html

Chapter 14 Practice Problems

Site	1	2	3	4	5	6	7	8
Calcium	0.8	0.65	40.7	1.68	14	22	241	400
Magnesium	1.2	0.14	7.2	0.24	13	17	7200	1350
Sodium	9.4	0.56	1.4	0.16	8	14	83,600	10,500
Potassium	-	0.11	1.2	0.31	-	0.5	4070	380
Bicarbonate	4	-	114	5.4	104	129	251	28
Sulfate	7.6	2.2	36	1.3	4.7	1.3	16,400	185
Chloride	17	0.57	1.1	0.06	8.5	33	140,000	19,000
Silica	0.3	-	3.7	0.7	24	30	48	3

These values are in mg/L.

1. Calculate the TDS for each water.
2. Plot its composition on a Piper Diagram (google search for these, or go to http://warmada.staff.ugm.ac.id/Graphics/gnuplot/diagram/piper.html)

Chapter 15. Biogeochemistry

15.1 Biogeochemistry

Life permeates the surface of the earth. It hugely influences all surficial geochemistry. In fact, most low temperature geochemical processes are probably biologically mediated. Life is an important part of rock weathering, controls most redox reactions in the subsurface, causes the formation of rock (most carbonate rocks are formed by life), and controls the composition of the earth's atmosphere. As such, understanding the varieties of life present on the earth, what life does to geochemistry, and how life makes economically important deposits is a crucial part to geochemistry.

15.2 Biochemistry in one subsection

There are whole majors devoted to biochemistry. It is not a small subject. We cannot do the subject justice in one subsection. That said, you can always take a course in biochemistry if that floats your boat. What follows are some of the tools that have been used to understand life's diversity, evolution, and biochemical processes that drive the local environment.

If you remember your high school biology (and never took a college biology course, like your author) you may remember that life is grouped into a hierarchical structure. Every organism belongs to a specific kingdom, phylum, order, family, genus and species. A species is defined as a naturally inter-breeding organismal population capable of producing fertile young. Groups of related species are tied to genera, which group into families, then to orders, then to classes, then to phyla, then to kingdoms. For instance, humans are in the animal kingdom, chordate phylum, mammal class, primate order, family hominidae, genus *Homo*, with the species *sapiens*. In many ways this separation is artificial, and not always quantitative.

The advent of biochemistry in the middle of the 20th century allowed scientists to determine biological and evolutionary relationships much more clearly. This is because all life is built around a central biochemical paradigm: in all life DNA codes for RNA that in turn codes for proteins. DNA, or deoxyribonucleic acid, is the material that is inherited from one generation of organisms to the next, either through cell division, sex, or one of the many other ways organisms pass along their genes. It consists of a set of organic compounds known as nucleotides that bond by hydrogen-bonding to other nucleotides to make a double helix. There are four nucleobases in DNA (A, G, C, and T); these bases bear the genetic information of life. Proteins include a large variety of materials including enzymes (biological catalysts) and structural material such as chitin. Proteins are composed of amino acids, of which there are 20 that regularly appear in proteins. RNA is ribonucleic acid and is like DNA in composition except it has an extra oxygen in its backbone, and replaces the T with U. The presence of RNA in this sequence points to the origins of the ancestor of life. Prior to the development of modern life (DNA→RNA→Protein), a simpler life may have been around that used RNA primarily. This sequence in life's early evolution is known as the *RNA World Hypothesis*.

You may notice that if life uses four nucleobases in DNA to define 20 amino acids in proteins. This difference demonstrates that the formation of proteins from DNA requires more than one nucleobase to code for a single amino acid. Indeed, DNA is composed of *codons*, sets of 3 nucleobases that correlate to a specific amino acid. Life has a significant amount of redundancy and error-tolerance in this process, and hence the DNA sequence of an organism can be used to determine its evolutionary relationship.

Figure 15.1. A fragment of RNA, DNA, and an amino acid. R can be replaced with one of the genetic letters (A, G, C or U for RNA, or A, G, C, and T for DNA), or any of several organic groups for the amino acid. Vertices on these diagrams are carbons bound to an appropriate number of hydrogens (take organic chemistry 1 for details).

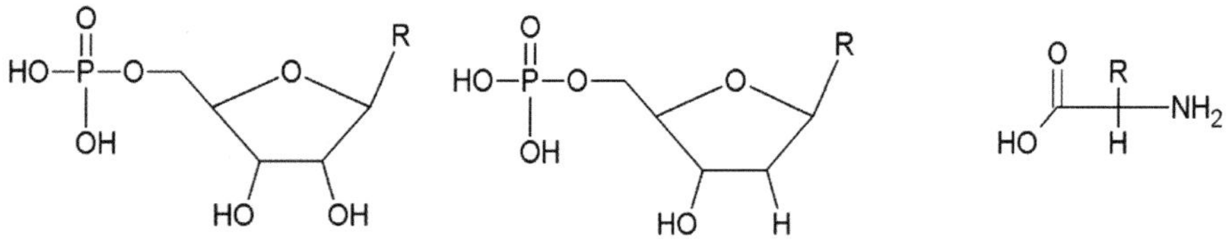

14.3 The Tree of Life and the dominant life-forms

Given there is a significant amount of redundancy implicit in the structure of DNA, there also is a significant amount of tolerance built in for error. In other words, if a DNA sequence is changed by a small portion, the protein the DNA makes usually is unaffected in its gross chemical properties. Hence, small variations occur at the mutation rate of an organismal population, and can be used to determine evolutionary relationships. For instance, if you have three sequences of DNA:

AGGTCCATATCGGCGATT

AGGTCCATATCGGCGATA

ACGTGCATAACGGCGATT

You can infer that the first and second sequences are more closely related to each other than they are to the third, based on the number of differences between these sequences. An analogy to this process is if three students cheat on a paper, and turn in their papers to be graded. If two of the papers vary only by one word out thirty, and the third paper varies from the other two papers by one word out of ten, then the first and second students cheated *more* than the third student (though all still get F's). There are similar approaches taken in linguistics (the study of language) that also show the relationships between languages. For instance, the following phrases are the provided in different languages (grabbed off the google translator).

The cat in the hat

Die Katze im Hut

Le chat dans le chapeau

Il gatto nel cappello

El gato en el sombrero

Kissa on hattu

Bu şapka içinde kedi

帽子の猫

You can tell that "Le chat dans le chapeau", "Il gatto nel cappello", and "El gato en el sombrero" are all closely related as they vary only in letter order and spelling. Indeed, these three (French, Italian, Spanish) are part of the Romance language family. Similarly, other languages are less closely related to these three, including Finnish (Kissa on hattu) and Turkish (Bu şapka içinde kedi).

One of the greatest advances of molecular biology in the past has been the discernment of the relatedness of life from the DNA (or RNA, or protein) sequences of single organisms. The degree of similarity can be organized into a diagram known as a phylogenetic tree (Figure 15.2). The awesome result of this analysis is that life groups into three *domains*. Indeed, things that seem to be quite different are actually close neighbors on this tree. Homo, for instance, is us humans, and Zea is corn. Corn is closely related to us on this tree. This diagram showed us scientists that most of the diversity of life is at the microbial level. There are three domains of life that include Bacteria, Archea, and Eukaryotes, and these divisions define most of the variety of life. One intriguing result from these studies are that mitochondria (which live in your cells) and chloroplasts (which make plants green) are actually bacteria that were absorbed by larger cells where they persist today. This grouping also showed us that the "kingdoms" of the 1960s' biology class (still taught today, naturally) were artificial and inadequate for describing life.

Figure 15.2. Schematic tree of life. Distances between points are roughly inversely proportional to the degree of relatedness.

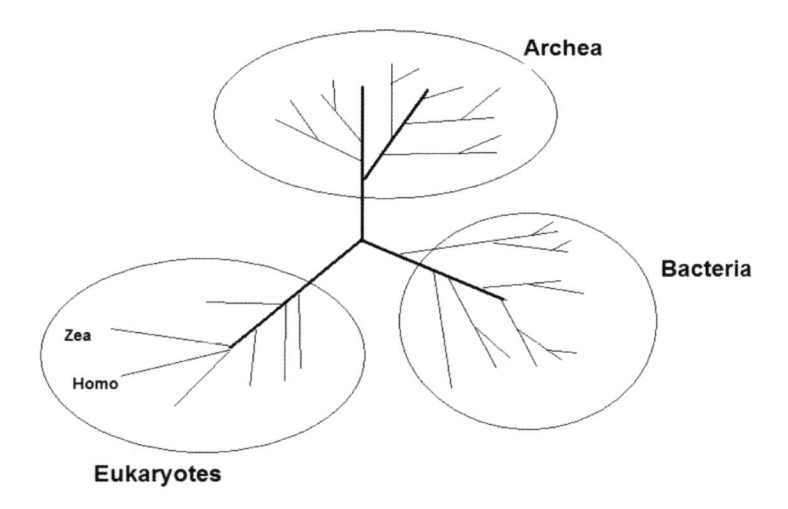

Nearly all life, as you can see from the tree, is microbial. Microbes are the dominant life form on the planet, and they control the geochemistry of many terrestrial systems.

15.4 How Life Affects Geochemistry

Most life is microbial. Microbes drive geochemistry on a local and often global scale. The next few chapters will deal with the role of life in biogeochemical cycles, as well as the role of life in organic production and rock formation.

Life can be divided into varying groups based on two features: what they use to build organic compounds, and how they get their energy. If an organism uses CO_2 to build its organic compounds, it is called an *autotroph*. Plants are autotrophs, as they get their carbon from atmospheric CO_2 and fix this to form sugar and other organics. If an organism gets its carbon from other organic compounds (for instance, from sugars or proteins), then it is called a *heterotroph*. You and I are heterotrophs as we get our carbon from the food we eat.

Additionally, the source of energy for an organism determines what type of organism it is. If it gets its energy from the sun, it is called a *phototroph*. Again, plants are a good example. If they get their energy from changing the chemical composition of food, then they are called *chemotrophs* (with you and me again being an example). Furthermore, another class of organisms is the *lithotrophs*, which get their energy from the oxidation of inorganic material, such as ferrous iron or sulfides. A full description of an organism may be a photoautotroph, which gets its energy from sunlight and builds its organic compounds from atmospheric CO_2 (e.g., a houseplant), or a chemoheterotroph, which gets its energy and carbon from organic compounds (e.g., a human).

Lithotrophs are only found in as microbes, however they are hugely important in many geologic processes. For instance, acid-mine drainage occurs when surface of near-surface water flows through a sulfide rich ore deposit, and becomes strongly acidic (pH ~2). This results from microbes oxidizing the sulfur in sulfide with atmospheric O_2 to generate lots of acid.

Biological signatures control the isotopic composition of environments. Most organisms have carbon in their bodies that are low in ^{13}C ($\delta\ ^{13}C$ < -20 per mil). Organisms also alter the sulfur isotopes of rocks, change nitrogen isotopes, and change hydrogen/deuterium ratios. Isotopic signatures are one of the diagnostic tools for life identification, though this process can provide ambiguous results.

One of the most important ways that life has influenced geochemistry on a global scale is in the atmospheric production of abundant molecular oxygen (O_2). Atmospheric O_2 is a waste product of photosynthesis. It is also a powerful oxidant. Photosynthetic microbes began the large-scale release of O_2 into the atmosphere sometime in the Archean (2.5 billion years ago and earlier). This O_2 forced a significant change in the surface chemistry of the earth, oxidizing iron from +2 to +3, making sulfides less stable, changing the chemistry of uranium, and many other things. Many of the minerals you see on the surface of the earth today would never have appeared if it wasn't for this change.

15.5 Redox and life

In addition to the origin of O_2, life is important in controlling the redox chemistry of local groundwater environments, specifically wetlands. If you imagine a typical swamp, you tend to think of mosquitoes, tall grass, smelly water, mud and muck, and perhaps leeches. Wetlands are among the most diverse and productive environments on earth. They are diverse due to the large range of redox environments present within their system.

Life gets its energy from the oxidation of sugar. A sugar is a compound with the nominal formula $C_nH_{2n}O_n$ (which, if you rewrite as n × C × H_2O, it should become obvious why sugars are called "carbohydrates"). A common sugar used by life is glucose, or $C_6H_{12}O_6$. Glucose is burned by organisms to generate CO_2 and energy. Since the oxidation state of C in a sugar is nominally 0 (check for yourself!), the sugar must lose a total of 24 electrons per sugar molecule to generate energy. If the carbon is being oxidized, something else must be reduced.

For aerobic organisms, the object being reduced is atmospheric O_2. In this chemical reaction, organisms turn glucose and O_2 to CO_2 and water:

$$C_6H_{12}O_6 + 6\ O_2 \rightarrow 6\ CO_2 + 6\ H_2O$$

Molecular oxygen is the only thing that larger organisms, such as you and your cat, can use to dump the extra electrons. Microbes aren't so limited.

In the wetlands environment, the amount of oxygen in the water is controlled by the slow diffusion of oxygenated water to the subsurface. In the case of the swamps, this diffusion ends at the water-soil interface (or close to it). This is because there is typically so much organics at this layer that further delivery of oxygen is buffered by the oxidation of decaying organic detritus. Immediately below the surface is an oxygen-poor environment that is not conducive to aerobic life.

Microbes have a variety of methods available to oxidize sugars even in these environments. The first material to be oxidized after O_2 is nitrate (NO_3^-). Nitrate is reduced to N_2 or to NH_4^+, depending on the organism and the environment. This redox reaction frees up slightly less energy than using O_2, but still plenty for most organisms.

$$C_6H_{12}O_6 + 3\,NO_3^- + 6\,H^+ \rightarrow 6\,CO_2 + 3\,H_2O + 3\,NH_4^+$$

After using up the nitrate, organisms use ferric iron as an oxidant (turning it to Fe^{2+}) and MnO_2, turning it to Mn^{2+}. Iron reduction is an important part of wetlands chemistry, and is easily apparent as red (Fe^{3+}) soils that become flooded turn to black as the iron is reduced.

Many microbes can tolerate using O_2, nitrate, ferric iron, and Mn^{4+} as electron receptors. These microbes are called facultative anaerobes, in that they can tolerate environments that are reducing enough to promote reduction of iron. However, once all the ferric iron runs out (or Mn^{4+} in those rare cases), their biologic activity stops. Other microbes start to act in those cases, and these organisms are called anaerobes.

Anaerobic respiration consists of two main redox reactions. The first is the reduction of sulfate (SO_4^{2-}) to sulfide (typically H_2S). This process is the one that results in the "rotten egg" smell characteristic of swamp muck. After reduction of sulfate to sulfide, anaerobic organisms will use dissolved carbonate as the electron acceptor, generating methane (CH_4) as the final product. This reaction is known as methanogenesis, and is one of the ways of producing green electricity from waste.

$$C_6H_{12}O_6 + 3\,HCO_3^- + 3\,H^+ \rightarrow 6\,CO_2 + 3\,H_2O + 3\,CH_4$$

A number of elements may also participate in these redox reactions, including several transition metals, arsenic, and possibly phosphorus. These reactions are quite important for environmental remediation, especially in the realm of organic contaminant cleanup. If a gasoline tank at a gas station leaks gas into the aquifer, cleanup sometimes includes supplying microorganisms with O_2 (by blowing it into the aquifer using a well and large fan), allowing them to oxidize the organics.

Chapter 15 Practice Problems

1) If a codon consists of three sequential nucleobases (A, G, C, T), to make one of 20 amino acids, how much redundancy is in the genome of organisms? That is, how many sets of 3 nucleobases are there? If these code for 20 amino acids, what would a single change to this sequence (for instance, a G turns into a C) do? Note that these codons code for 21 things to make a protein: the 20 amino acids, and a code to tell the synthesis of the protein to stop.

2) Iron pyrite in mine tailings is one of the nastiest mineral sources of acid-mine drainage. In this reaction, FeS_2 (pyrite) reacts with O_2 from the air in water to generate H^+, Fe^{3+} and SO_4^{2-}. Balance this redox reaction.

3) Balance the redox reactions that involve glucose and Fe^{3+}, and glucose and SO_4^{2-}, e.g. $C_6H_{12}O_6 + Fe_2O_3 \rightarrow CO_2 + Fe^{2+}$, and $C_6H_{12}O_6 + SO_4^{2-} \rightarrow CO_2 + H_2S$.

Chapter 16. Nutrient Cycles

16.1 What is a nutrient?

Nutrients have an intuitive meaning, and indicate the "healthy-ness" of a material. On a more fundamental level, nutrients are those materials that cannot be made by an organism that are necessary to its survival. All nutrients are determined by the local geochemical environment. Nutrients are usually limited in availability either due to rarity or inaccessibility; these two are not necessarily the same. Nutrients are usually specific materials that can do a specific chemistry or chemical reaction that no other compound can do.

All organisms that we know of need water, energy, carbon, nitrogen, phosphorus, sulfur, salts, and metals to live. If one of these is limited in availability in an environment, it is said to be the limiting nutrient of that environment. In most cases, nitrogen, phosphorus, salts, or metals are the limiting nutrient in the terrestrial environment. Other nutrients are a bit more specific: vitamins, for instance, are unique to specific organisms.

In general, a vitamin is a complex organic compound that plays a specific role in biochemistry. Vitamins get their name from "vital amine", with amine being an organic group bound to a –NH_2 group. There are several vitamins (check out your average multivitamin), but most are accessible with a varied diet. Vitamins are in many cases the result of evolutionary adaptation. For instance, vitamin C, which is found in fruits and vegetables, is a necessary vitamin that prevents scurvy. However, vitamin C is a vitamin only for primates. Other animals, such as dogs and cats, can make their own vitamin C. However, our fruit-eating ancestors had an abundance of vitamin C in their diets, and hence when the genetic machinery for vitamin C construction broke down millions of years ago, this was no cause for concern (unless traveling on the open seas without a supply of fruit!). In contrast, cats cannot make the amino acid arginine, as they get it from the meat they ate. As a result, housecats are often arginine-deficient. Adult humans can make their own arginine, as this has been a necessary part of our biochemistry.

The case of vitamins being unique to specific groups of organisms breaks down at the elemental level, as organisms cannot do nuclear fusion. The geochemical cycling of carbon, nitrogen, phosphorus, and metals is an important part of biochemistry, and has been since the first organisms arose on Earth.

16.2 Carbon Cycling

A reactive source of carbon is viewed as one of the fundamental requirements for the origin of life. After life originated on the earth, carbon was rarely a nutrient, as in most cases there is enough carbon as CO_2 to supply carbon for most major organisms. Carbon on the earth lies in two major forms: oxidized carbon as atmospheric CO_2 and carbonate rock (limestone), and reduced carbon as biological material and former biological material (e.g., coal and oil). The transformation of oxidized carbon to reduced carbon takes place primarily through *primary production*, which includes photosynthesis. Most of the carbon formed this way ends up being oxidized by organisms for fuel back to CO_2. However, some portion of this carbon can be buried, where it eventually becomes fossil fuels such as oil and coal. Anthropogenic activities recently have been oxidizing the large reservoirs of reduced carbon from underground to power our modern industrial society. As a result, the carbon cycle is shifting to more oxidized varieties. Most of the earth's carbon is stored in carbonate rock.

16.3 Nitrogen Cycling

In most cases carbon is plentiful enough that it is never a limiting nutrient. In oceanic and other environments, the elements nitrogen and/or phosphorus (and sometimes iron) are the limiting nutrients. Nitrogen may seem surprising, since it makes up most of the atmosphere, with 78% of the molecules you are breathing right now consisting of N_2. However, nitrogen is limiting due to its chemical structure. The nitrogen atoms of nitrogen gas (N_2) are bound together by a triple bond (Figure 16.1). Triple bonds, as you might suspect, are much harder to break than single bonds, and hence N_2 in the air does not readily enter biomolecules. Nitrogen is used extensively by life in several biomolecules, including amino acids, which make up proteins, and nucleobases, which are the letters of DNA.

Figure 16.1 Structure of dinitrogen.

$$:N{\equiv}N:$$

The process of turning nitrogen in the atmosphere (N_2) into nitrogen that can react to form organic molecules is called *nitrogen fixation*. Nitrogen fixation is either the oxidation of N_2 to form NO-type molecules (such as NO and NO_2), or the reduction of N_2 to ammonium and ammonia (NH_4^+ and NH_3). In most cases, once the N_2 bond is broken to form either NO or NH bonds, biogeochemical processes will readily change the redox state to suit the needs of the local environmental community (in other words, it doesn't matter as much if the nitrogen is oxidized or reduced, just that it's not triply bonded to itself). Nitrogen fixation is a process performed by a small number of microorganisms, and is an energy-intensive process. In fact, humans have surpassed biology in nitrogen-fixing potential: humans now fix more nitrogen from the air each year than do microorganisms, largely to feed the human population. Since nitrogen fixation is a high-energy process, there are relatively few ways to fix N₂ by abiotic pathways. Lightning is one pathway, as the high-energy electric discharge breaks apart N₂ molecules, allowing it to recombine randomly with O atoms, forming NO atoms. Other high energy processes (volcanic eruptions, meteorite impacts) may also fix nitrogen.

Figure 16.2 DNA molecule.

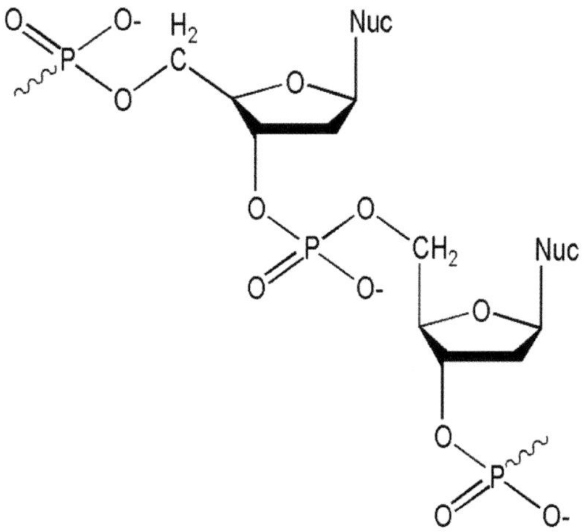

16.4 P cycling

Phosphorus, as phosphate, has a few key characteristics that make it ideal for metabolic processes and for the construction of biomolecules. Phosphate resides primarily in four reservoirs in cells: as free phosphate within cellular plasma, in membranes as phospholipids, in the nucleic acids (Figure 16.2), and as metabolic molecules including ATP. Phosphate is so useful because of its ability to bond to two materials yet retain a negative charge. Charged molecules are more soluble than uncharged molecules in water, yet less soluble in organic membranes, hence life sticks phosphate onto molecules extensively to keep them in place in cell material.

Phosphorus is a nutrient primarily because phosphorus follows the rock cycle. Unlike the other nutrients we have discussed (C and N), there is no significant volatile phase for phosphorus. Phosphorus is almost exclusively in phosphate, and is usually found in mineral form in rocks. Phosphate minerals in rocks are sparingly soluble, and hence the slow leaching of phosphate from rocks is the only way of liberating phosphorus for use in biology. In general, phosphorus is typically only in phosphate (+5 oxidation state) as well. The other major biogeochemical elements, and most metals have much greater variety in their redox state. The features of the phosphate cycle include low solubility, slow cycling, and limited redox evolution, making phosphorus a critical element in many ecosystems.

16.5 Metals

Life is typically characterized as "CHNOPS" elements (carbon, hydrogen, nitrogen, oxygen, phosphorus, and sulfur, and pronounced like the alcoholic beverage), and these elements together make up about 99% of the mass of the average organism. However, the role of that last 1% is not insignificant, and consists mostly of metals. These metals include monovalent salts such as Na and K, as well as divalent Mg and Ca, and true metals such as V, Fe, Cr, Mo, and even W. Many of these metals are used because they catalyze specific chemical reactions (such as delivering oxygen to cells as Fe^{3+} and Fe^{2+} in heme, or splitting dinitrogen using Mo). These elements may not be needed in huge quantities, but are critical in small amounts, and life may not live at all in environments where they are not present. Intriguingly, the composition of metals across many organismal groups is quite similar. This points to some antiquity for the metallic composition of life.

We have reason to suspect that life's current metal composition is ancient. This idea has come about due to studies of how elements likely behaved on the early earth, and how these behaviors compare to modern biological composition. If the modern-day elemental composition of the ocean is compared to the modern-day composition of cells, there is a decent correlation between the two (see schematic figure 16.3), though there is significant scatter. However, since life has been modifying the geochemistry of the earth for the last 3.5 billion years at least, using the modern ocean composition does not present an accurate picture of the environment in which cells arose. The early earth, lacking oxygen, was influenced by a reducing environment, rich in free electrons, as evidenced by the presence of pyrite and uraninite minerals on the early earth's surface. Both minerals react away in an oxidizing environment. If the composition of the ocean is changed to account for a reducing environment, the relationship between modern cells and the ancient ocean significantly improves. The composition of living cells reflects in part the environment in which those cells evolved, and hasn't changed too much in the last few billion years. For more discussion on early geochemistry relationships with stoichiometry see Byrne (2002).

Figure 16.3. Schematic of the elemental abundances of human blood serum (used as a proxy for general biochemical systems), compared to the elemental abundances of two oceans, with A) showing the modern ocean and B) showing a reducing ocean.

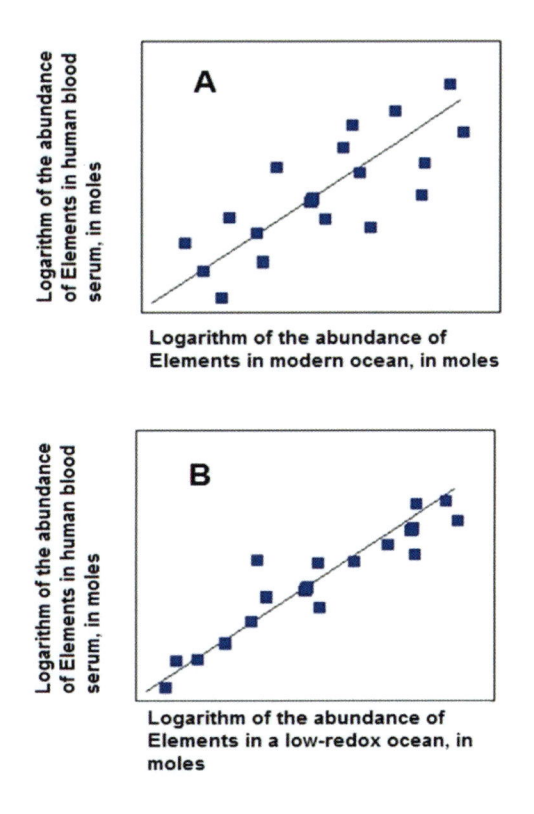

16.6. Synthesis

Life requires water, carbon, nitrogen, phosphorus, and metals as its elemental/molecular inventory. These elements are necessary because they perform some sort of critical biochemical process. Many of these requirements are deeply rooted: the earliest materials available have also become the most critical for life's development. Most of these elements cycle though biological processes, and are strongly influenced by geochemical availability.

The geochemical and environmental availability of nutrients impacts biological productivity. In no way is this made clearer than what happens there is a sudden influx of a previously limited nutrient. For instance, if phosphate (from lawn fertilizer, perhaps) is washed into a nearby pond during heavy rain, algae within the pond have just received a significant boon. The algae respond by growing to consume the phosphate. Surface algae subsequently expand to cover the surface of the pond, blocking light from reaching the subsurface, and starving the subsurface of oxygen. The subsurface becomes anoxic, and results in the *eutrophication* of the water body (see Figure 16.4).

Figure 16.4. Eutrophic pond covered with algae. See author for scale.

Chapter 16 Practice Problems

1. The ratio of C:N:P in modern oceans follows a stoichiometry called the "Redfield Ratio". Throughout most of the ocean the C:N:P is 106:16:1 in terms of atoms to atoms. If an element is less than this ratio, then it is said to be limiting. If you measure a carbon content of 50 mg/L, a nitrogen content of 2.2 mg/L, and a phosphorus content of 0.15 mg/L, which of these elements is limiting?

2. Nitrogen as ammonia is capable of reducing nitrate to release N_2. Balance the equation:
$$NH_4^+ + NO_3^- \rightarrow N_2$$

3. Using the thermodynamic properties of several substances in the table below, estimate the solubility of the mineral apatite (in mg/L) in water at a pH of 8. Assume the dissolution reaction is:
$$Ca_5(PO_4)_3F + 3H^+ \rightarrow 5Ca^{2+} + 3HPO_4^{2-} + F^-$$

Species	ΔG^0 (kJ/mol)
$Ca_5(PO_4)_3F$	-6475.72
H^+	0
Ca^{2+}	-553.676
HPO_4^{2-}	-1093.67
F^-	-277.884

Chapter 17. Organic Geochemistry

17.1 Organic Molecules

Carbon is a critical element in life and in our economy. This is because carbon is the most important element in organic chemistry, which comprises both biological carbon molecules, and the molecules they change into as oil and coal. As with biochemistry, there are whole majors devoted to organic chemistry, and we will only cover the topic here as it relates to geologic systems, and even then, only a small portion of that.

Note that in the science sense, organic has nothing to do with "natural" or "from the earth". Sarin, DDT, and Agent Orange are all organic molecules. Indeed, most of organic chemistry deals with the chemical synthesis of novel organic molecules that would probably have never existed without intelligent intervention (a tangentially related pet peeve of mine is if you see any substance that is "chemical-free" it means you are literally buying nothing, or a vacuum. Shampoos and cosmetics are the biggest offenders here). An organic molecule necessarily has carbon as a constituent. However, not all carbon-bearing materials are organic. As an example, calcite is $CaCO_3$, but is not considered organic. Atmospheric CO_2 is not considered to be an organic molecule. Carbon in silicon carbide (SiC) is not considered to be an organic molecule. To be considered as an organic molecule, the carbon must be bound via covalent bonds to another carbon atom, or to hydrogen (with maybe a few exceptions to this definition). Calcite and CO_2 are not organic molecules since there are no C-C or C-H bonds, and the bonds in silicon carbide are closer to ionic bonds than they are to covalent bonds and hence this is not considered to be an organic molecule. Some scientists also don't consider gases such as HCN (hydrogen cyanide) or formaldehyde (CH_2O) to be organic, but for the purposes of this material, let's think of them as part of this subject area.

The naming conventions of organic compounds is somewhat complex and beyond the focus of this course, but the name of an organic molecule tells you something about how many carbon atoms are organized into a chain (Table 17.1).

Table 17.1. Carbon numbers and names

Carbons	Prefix
1	Meth-
2	Eth-
3	Prop-
4	But-
5	Pent-
6	Hex-
7	Hept-
8	Oct-
9	Non-
10	Dec-

Additionally, many organic molecules have different things bound to them, which results in them behaving differently with respect to their physiochemical properties. These changes are called variations in functional groups (Table 17.2) and alter the name of an organic compound. Note that a common symbol you will see in organic chemistry is "R". R represents a "generic" organic molecule part (short-hand for "radical"), and could be as simple as H or CH_3, but can include things with many more carbons than those.

Table 17.2. Functional Groups and names

Functional Group	Name
R-**OH**	Alcohol
R-**COOH**	Carboxylic acid
R-**CHO**	Aldehyde
-**C=C**-	Alkene
-**C≡C**-	Alkyne
R-**CO**-R	Ketone
R-**O**-R	Ether
R-**CO-O**-R	Ester
R-**NH₂**	Amine

17.2 Bonding structures in organic molecules

Atomic orbitals (those s, d, and sometimes f things from Gen Chem 2) of atoms can hybridize to make new orbitals. Carbon, specifically, can form sp, sp^2, and sp^3 orbitals (Figure 17.1). A given carbon atom can make either 2 sp orbitals, 3 sp^2 orbitals or 4 sp^3 orbitals. This determines the shape of the molecule. Molecules with sp orbitals are linear at the carbon with the sp bonds, whereas one with sp^2 orbitals end up planar with other atoms bound to it at 120° angles, and one with sp^3 bonds become tetrahedral with other atoms at about 109.5° angle between each other. An example of a sp bonded molecule is the carbon in acetylene (C_2H_2), of sp^2 hybridization is the carbon within benzene (C_6H_6), and with sp^3 hybridization is octane (C_8H_{18}), from gasoline.

Figure 17.1. Orbital diagram

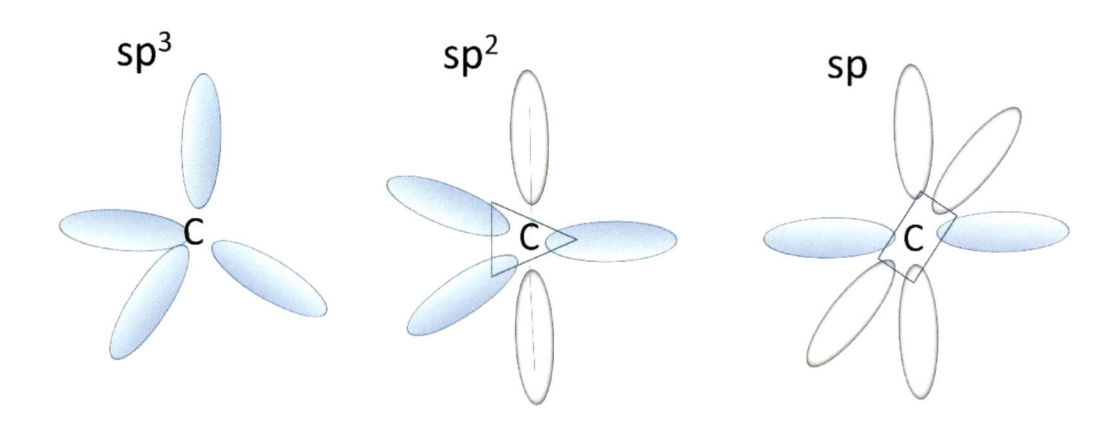

In general, the two most common hybridizations of carbon encountered in nature are sp^2 and sp^3. For a given organic compound occurring within a rock, the ratio of sp^2 to sp^3 hybridized carbons tells you something about its metamorphic history. Most organics from biological organisms and other processes start out sp^3 hybridized. However, as this carbon is heated, it slowly converts to sp^2 hybridization. This process is called *maturation* when it deals with organic compounds. In addition to changing carbon hybridization, the organic compound also begins to lose H atoms, and as a result, has a changing C/H ratio. For sp^3 organic compounds, this ratio is less than 1, but as the organic compound matures and gains more sp^2 hybridization, the ratio becomes greater than or equal to 1. Organic compounds characterized by mostly sp^3 hybridized bonds are called *aliphatics*, and organic compounds characterized by mostly sp^2 hybridized bonds are called *aromatics*.

Organic molecules can linear or branched. Linear molecules make longer chains, whereas branched molecules tend to be clumpier (Figure 17.2). These characteristics can influence the properties of an organic molecule. For instance, a molecule with a formula C$_8$H$_{18}$ can be linear, as octane, or can be quite clumpy, such as 2,3,3-trimethylpentane. Octane boils at ~125°C, whereas 2,3,3-trimethylpentane boils at about ~115°C, despite having the same mass.

Figure 17.2. Two molecules with the formula C_8H_{18}, having different structures and properties.

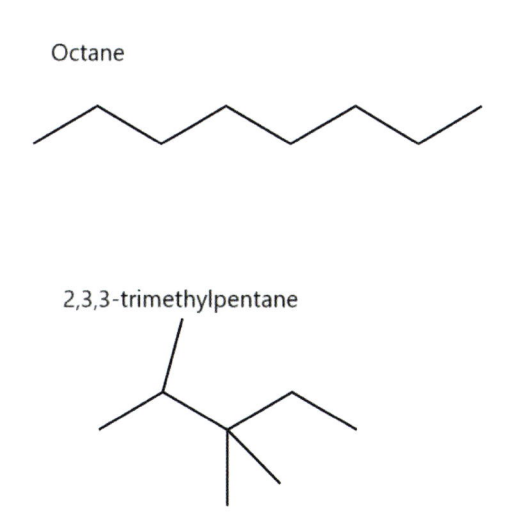

17.3 Petroleum

Our modern society runs on the organic compounds produced by microbes millions of years ago. Most organisms are part organic and part water, and, when buried, these compounds separate during diagenesis. Organic compounds are typically less dense than water and do not mix well, hence they will tend to float on top of water. As a sediment rich in organic material is heated (for instance, by burial and compaction), volatile or light material will float to the surface on top of water. Most petroleum is collected at the top of an anticline capped by an impermeable layer (Figure 17.3).

Figure 17.3. Petroleum can be trapped at the top of an anticline beneath an impermeable layer as oil is less dense than water.

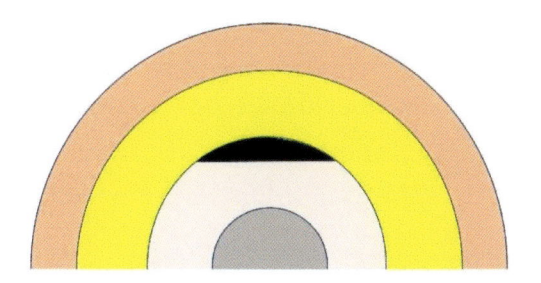

Petroleum has four major subclasses of material that are used for different purposes. *Paraffins* are small aliphatic compounds (usually 10 carbon and lighter) that are typically straight-chained. These are among the highest in energy for burning. Gasoline for cars is part of this material. *Naphthenes* are also dominated by aliphatic compounds, though many are slightly larger and less linear than paraffins (including branched organics and ring compounds), and these provide slightly less energy for burning. Both naphthenes and paraffins are composed of sp^3 carbons. *Aromatics* are compounds that are mostly sp^2 hybridized and hence are poorly burning. *Asphaltics* are highly polymerized (molecules that are all interconnected) aromatic materials and are largely solid and low-energy for burning.

The quality of an oil deposit depends on the relative ratios of these four major subclasses. Oil with mostly paraffin and naphthene compounds is more valuable than aromatic- and asphaltic-dominated oil. The former is more liquid, and has more energy available on burning. It also burns cleaner, with less soot released. Much of the soot that is released during burning consists of aromatic compounds. Oil dominated by asphaltic molecules is better known as *tar* and is used primarily as a glue to bind rocks together in asphalt (hence the name).

Petroleum matures because of metamorphism and oxidation. Carbon molecules, as they are heated or oxidized, release hydrogen or water and create more carbon-carbon bonds. As a result, the C:H ratio becomes larger, and the hybridization starts to favor sp^2.

Oil's origin is well established. Although some individuals in the political realm may say otherwise, oil is formed from the decay of phytoplankton and zooplankton in a shallow marine environment. There are a few ways we know this. For one, some of the molecules within oil match the organics used by plankton today. Some molecules are unique to these plankton; many of these are steroids. Within petroleum are several molecules that match these steroids, strongly pointing to an origin from these creatures. Additionally, these plankton use certain rare metals, in this case nickel and vanadium, in their biochemistry. Petroleum, interestingly enough, has both Ni and V in a ratio close to the ratio found in plankton.

There are a few other synthetic pathways that produce oil. Oil can be formed by the corrosion of carbon-rich metal (such as carbon dissolved in iron). In fact, this was an idea first proposed by Mendeleev, the father of the periodic table of the elements. However, since metallic iron is so rare on the earth, this pathway is unlikely to be very important. Other natural phenomena that might produce oil include serpentinization as mentioned in the metasomatism chapter, which releases reduced gases, primarily H_2. However, if this H_2 is around CO_2, a reduction reaction can occur:

$$CO_2 + 4\,H_2 \rightarrow CH_4 + 2\,H_2O$$

This process may in fact be the origin of Martian methane. This methane itself can react to generate larger organic molecules, though at decreasing abundances.

17.4. Coal

Coal is a macromolecule. A macromolecule is a large molecule. Coal consists of multiple interconnected carbon and hydrogen (and a few other elements) that make a disordered, unstructured mess. If you hold some coal, you are effectively holding a single molecule (Figure 17.4). Due to this highly networked nature, coal is a solid, though it is similar in composition to oil or petroleum. Like oil, coal is affected by heating and "matures" by decreasing its C to H ratio, and increasing the percentage of sp^2 hybridized carbon.

Figure 17.4. Schematic of coal's molecular structure.

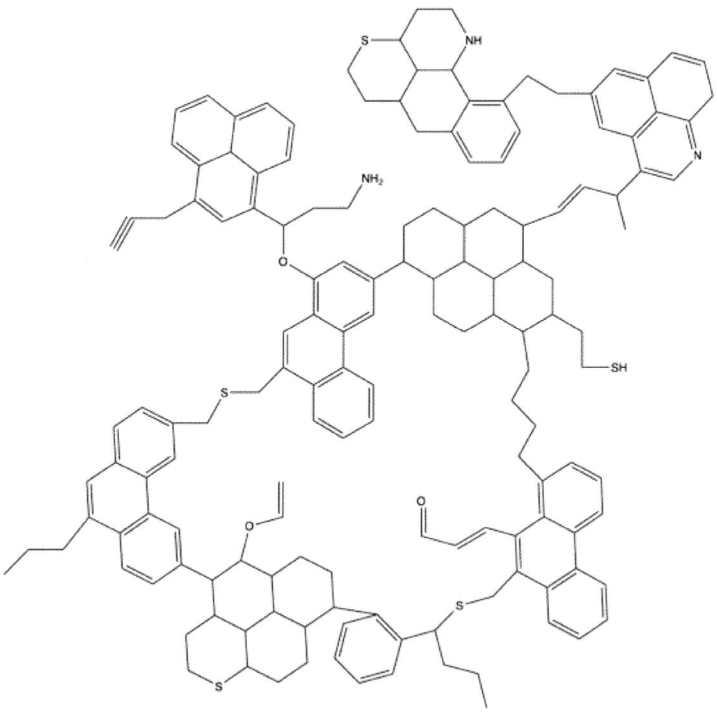

There are broadly four types of coal or coal precursors. *Peat* is the starting material of coal. Peat consists of plant matter that has accumulated from rotting plants, often in a wetlands environment. It is usually quite full of water at the molecular level, but it has been used historically for many years as a source of energy. Of all coals, it is the closest to a renewable energy source.

If peat is subjected to some heating and pressure, then some of the water is released from the organic mix, drying it and resulting in a precursor of coal called *lignite*. Lignite is the lowest grade of coal, though it can also be considered a high grade of peat. When lignite is turned into rock by diagenetic processes, it becomes a coal called *bituminous coal*. This is the black coal that is used as a threat to misbehaving children around Christmas time. If bituminous coal is metamorphosed, it becomes a variety of coal known as *anthracite*. Anthracite is a shiny coal that often has a metallic luster. If this material continues to undergo metamorphism, it eventually becomes pure carbon as *graphite*, or under high pressure, *diamond*. The formation of graphite in anthracite is part of what makes this coal silvery and shiny.

In general, the energy released by combustion (burning) of an organic material is proportional to the amount of sp^3 hybridized carbon in it. sp^2 carbon has less energy and tends to provide much less energy per mass burned into CO_2. Peat has the most sp^3 carbon, whereas anthracite has the least. With this in mind, it would seem as though peat is the best energy source. However, peat is also the wettest of the coal varieties. When it is burned, much of the energy stored in the molecules of peat goes into vaporizing water. Hence this material is much less energy dense than lignite and bituminous coal (which has the most net energy potential) and anthracite coal (Table 17.3).

Table 17.3. Energy content of coal

Coal Type	Energy (MJ/kg)
Peat	2-10
Lignite	10-20
Bitumen	28-35
Anthracite	26-33

17.5 Remediation

The modern use of petroleum as our primary source of energy has led to the formation of multiple purveyors of this fuel source. You know them better as gas stations. Gas stations frequently leak. Gas enters the surficial aquifer, where it acts as a poison for the local environment. This often then requires the hiring of environmental scientists and geologists to remediate the aquifer by removing the petroleum.

There are many varieties of organic pollutant that can require remediation aside from petroleum and petroleum products. Solvents, dry cleaning fluid, insecticides and pesticides, all can enter the aquifer from leaky storage containers and wreak havoc. Organic pollutants can be broadly classified into light non-aqueous phase liquids (LNAPLs) and dense non-aqueous phase liquids (DNAPLs). LNAPLs float on water whereas DNAPLs sink. Remediating the two is a whole course of its own, and will not be covered in depth. Most petroleum spills are LNAPLs.

In general, remediation of organic pollutants takes place either *in situ*, within the aquifer, or *ex situ*, by extracting the pollutant and cleaning it up. Ex situ cleaning can include 1) digging it out from the ground (if the spill hasn't sunk too far), and burning the contaminant, 2) pumping air into the aquifer and having the pollutant volatilize and then burning it, or 3) pumping it out and recovering it. In situ cleaning 1) includes pumping air into the aquifer and having the air oxidize the pollutant, 2) planting trees to uptake the contaminated and burning the leaves, or 3) adding an oxidizing compound to the aquifer to promote oxidation in situ. Sometimes cleaning does not work, and the pollutant is instead contained by blocking its movement with an impermeable wall.

In many cases, nature will do the hard work. Microbes will frequently use organic pollutants as energy sources and oxidize these compounds with O_2 or other oxidants and get rid of the problem. Such a process is called *natural attenuation* and is the cheapest cleanup possible, as the only cost is monitoring the pollutant plume.

17.6. Extraterrestrial Organics

We now move to something completely different. Some varieties of meteorites are loaded with organic material. Many of these organic compounds include things that are used in life today such as amino acids and sugars. The presence of these biochemically useful molecules in meteorites has led some scientists to argue that meteorites provided the building blocks of life.

Many meteorites contain carbon. This carbon is frequently in carbonates, or dissolved as an alloy in metal, or in diamonds, or in rare carbide minerals such as SiC. In other meteorites, the carbon may be in organic molecules. In general, this organic carbon is either in a water-soluble form, or in a macromolecular form like coal. Most of the time it's in the macromolecular form (between 50-100% of carbon is insoluble). However, the compounds that are soluble are the most interesting.

How do we know that these organic compounds were formed in space? In many cases, the instant a meteorite lands on the earth, it becomes contaminated with terrestrial organisms, which strongly alter the organic chemistry of the meteorite. In some cases, the meteorite is collected quickly enough to minimize this contamination. Studies of these organic compounds have showed three main features that place their origin as extraterrestrial. 1) they are enriched in ^{13}C, D, and ^{15}N. These isotopes are rare on the earth, but are a fair bit more common in the outer, cold regions of space. 2) They possess both handedness for molecules. Organic molecules formed by organisms consist of only one hand (*chirality*), depending on the molecule. Meteorite organics do not show this preference. 3) They decrease in abundance with increasing carbon number. Since organic compounds are formed by chain building in non-living systems, to build a 3C molecule you must build it out of a 1C and 2C molecule. In general, there must be a lot of 1C and 2C around to build a little 3C. Then, if you want to build a 4C, you must have enough 3C and 1C, with the 3C even rarer. In contrast, organic compounds formed by life have certain key numbers of carbon, depending on the metabolic pathway that formed them.

Based on the characteristics of the organic material found in meteorites, scientists have determined two probable formation pathways. One is the formation of these organic compounds on ice grains in the cold reaches of space. This origin would explain the isotopic heaviness of these compounds. A second origin is the water-alteration of existing organic compounds to make new molecules. Water (as either liquid or steam) could have flushed through an asteroid, grabbing and tearing apart organic material to make new molecules.

Like petroleum and coal, we see changes to meteorite organics because of processing. Meteorites that are "pristine" have seen little heating and oxidation of their organics, resulting in sp^3 hybridized carbons dominating. Those that have been heated extensively are dominated by sp^2 hybridized carbon.

Chapter 17 Practice Problems

1. Both graphite and diamond are varieties of carbon that can result from the metamorphism of coal. How much energy can you extract from burning graphite (in MJ/kg)? How much energy from burning diamond (in MJ/kg)? Believe it or not, the burning of diamonds was how scientists showed that diamonds are indeed made of carbon.

2. Let's say you are hired to remediate a nasty spill consisting of a DNAPL with the molecular formula CH_2Cl_2. You decide to use potassium permanganate ($KMnO_4$) as an oxidant to remove this pollutant. The permanganate is reduced to form MnO_2, and oxidizes the CH_2Cl_2 to CO_2 and Cl^-. Balance this redox reaction. If the $KMnO_4$ costs $1/kg, how much would it cost to clean 1 ton (1000 kg) of the DNAPL?

3. Identify the origin of these two organic samples as either meteoritic or terrestrial in origin. The X-axis is carbon number (1 carbon, 2 carbon molecules, etc.) and the y-axis is abundance of compounds with those numbers of carbon.

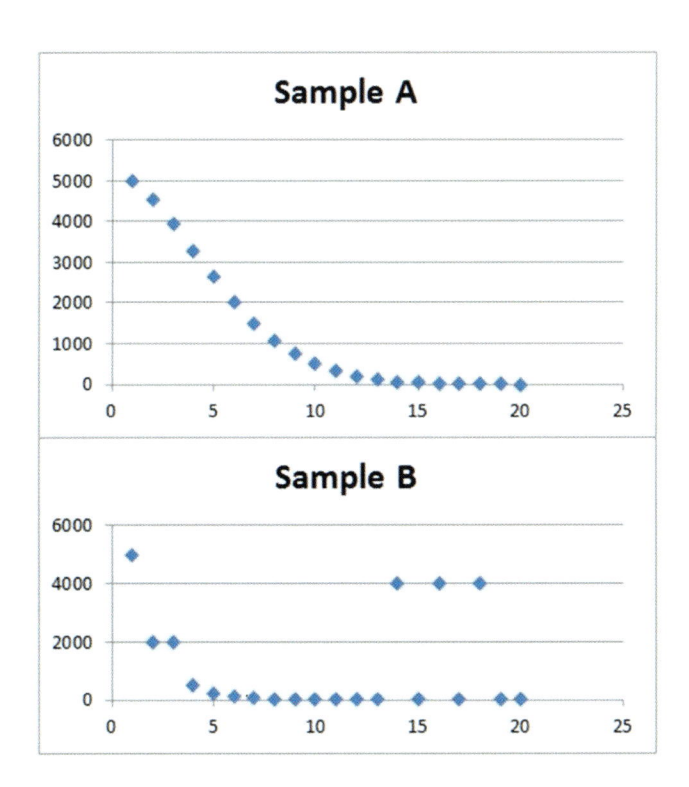

Chapter 18. Biomineralization

18.1 What are biominerals?

A biomineral is a mineral that is formed by a living organism. Biominerals are important economically, as many of these serve as crucial deposits of building material and fertilizer. They are formed by organisms for protection, eating and chewing, internal support, determining direction (north/south, or up/down), and sometimes as waste products. Several biominerals are found throughout the geologic record, and their presence is often indicative of the geochemical environment. Biominerals may be useful in the search for life outside of the earth as they differ in critical ways from abiotic minerals.

18.2 Examples of biominerals

There are several biominerals that are found throughout the geologic column. Probably the two most important are calcite and aragonite, as these two make up limestone, which one of the most common sedimentary rocks on earth. It is also important for building and construction, and has been used since antiquity for a variety of purposes. Calcium carbonate minerals are formed by many organisms, including mollusks, corals, echinoderms, and microbes. These minerals are formed for protecting organisms, and for structural support.

Figure 18.1. Cell precipitate biominerals, using organic matrices and proteins, and which form specific crystal structures. These evolve into larger crystals over time after the organism dies (artistic license taken to show the passage of time as they get larger).

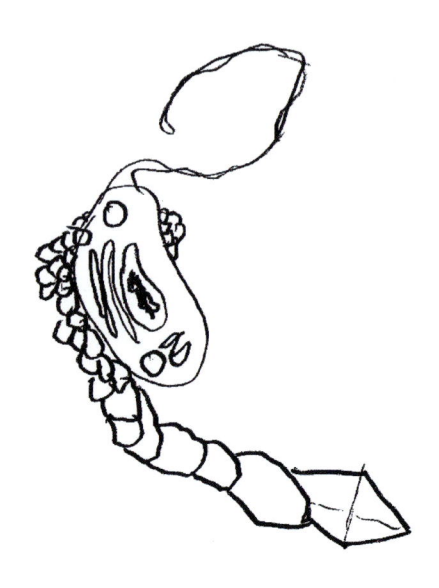

Vertebrates are dependent on apatite as a biomineral. This is the mineral that makes up bone. Bone gives us shape and protection. Apatite is also precipitated by some microbes, and deposits of apatite (known as phosphorite) likely have a biological origin. Phosphate rock is critical to modern agriculture. Other phosphate minerals are also produced by organisms unintentionally, such as struvite, which can form kidney stones in mammals.

Silica is formed by a variety of microbes, primarily diatoms, and it makes up diatomaceous earth. Silica is probably produced for protection by these organisms. Magnetite is used by some microbes and multicellular organisms for a variety of reasons, including magnetic orientation. Several other iron minerals are produced by unicellular organisms, as are some sulfides and sulfates. Some of these minerals are produced to determine which way is up by gravity.

18.3 How biominerals are formed

Biominerals are formed as cells excrete a mineral, typically externally. The mineral is precipitated on a protein template that organizes the atoms to form a mineral of a specific composition (Figure 18.1). The formation of biominerals is done by locally enhancing the activities of solvated ions (for instance, calcium and carbonate for aragonite precipitation), and typically does not occur well in undersaturated solutions.

Biominerals differ from terrestrial minerals in a few key ways. For one, biominerals are much less crystalline than abiotic minerals. This is because crystallization typically occurs at the cellular level, which means crystals merge during biomineralization, with multiple crystal domains. This is especially evident when viewed in thin section. Most biominerals are also very rich in organic material, often at 1-10% carbon by weight. Again, this is because the formation of biominerals occurs as proteins direct the formation of biominerals, and hence several of them can get caught during their formation. Although this poor crystallinity and organic contaminants would suggest that biominerals are brittle, the reverse is true. Biominerals are typically stronger than abiotic minerals. Finally, biominerals, since their formation is mediated by biology, form crystals much faster than abiotic processes.

18.4 Geochemistry of biominerals

The widespread geologic formation of biominerals can be traced to the origin of hard body parts. This occurred nominally in the Cambrian period, although precursors to hard body parts also appeared in the Ediacaran period. Microbes were producing biominerals well before this time. The origin of multicellularity accompanied by hard body parts is an area of active research. It is possible that the geochemistry of the time may have allowed for the mass production of biominerals, by enhanced weathering of continents due to a mild climate.

Biominerals are important in element cycling due to their solubility in undersaturated solutions. Within the modern ocean, there is a region below which calcite spontaneously dissolves away known as the calcite compensation depth. Below this depth, shells and calcite from microbes dissolves away rapidly as the concentrations of Ca^{2+} and CO_3^{2-} or carbonate are too low. This depth determines the extent that most invertebrate lifeforms (e.g., corals, shells, bryozoans, etc.) can live.

One of the intriguing properties of biominerals is that they can show how evolution was affected by local geochemistry. One specific way is the formation of either aragonite or calcite as a biomineral by organisms. In ocean water that is rich in calcium relative to magnesium (molar ratio Mg/Ca < ~4), calcite is easier to form than aragonite. In contrast, when ocean water is enriched in magnesium (Mg/Ca > ~ 4), aragonite is easier to form than calcite. This effect is not a thermodynamic effect but a kinetic effect. Magnesium interferes with the production of calcite, and hence aragonite is made in lieu of calcite. The chemistry of the ocean has changed throughout the phanerozoic, with the ratio of Mg to Ca changing over time. Organisms that began making hard parts chose to use either aragonite or calcite, depending on the composition of the ocean at the time (Figure 18.2). Even if the ocean chemistry changes, those organisms continued to make only the carbonate mineral the originally could make.

Figure 18.2. Origination time of major organism groups and sea chemistry. The minerals chosen by organisms to make their hard body parts reflects ocean chemistry at the time, which is either dominated by conditions conducive to calcite formation or aragonite formation. For instance, trilobites, originating in calcite seas, made calcite their biomineral (for their eyes, for instance).

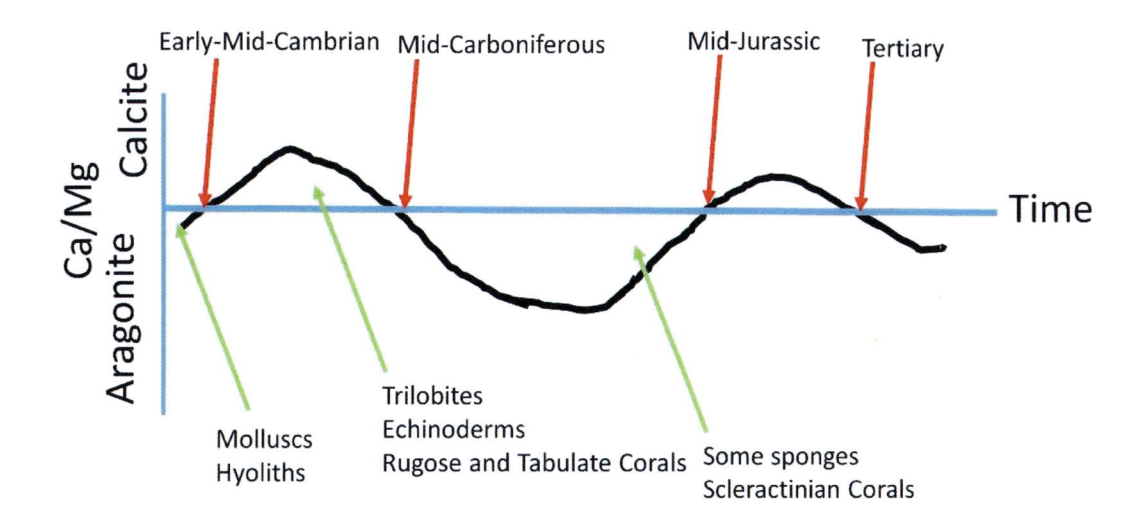

18.5 Biominerals as biomarkers

Biominerals can serve as biomarkers if an abiotic origin for a mineral can be excluded. These may be quite important if we ever try to find life on another planet. One example of biomarkers being extremely important in astrobiology is in the mineralogy of the Martian meteorite ALH84001 (these numbers designate the region of Antarctica where this meteorite was found—Allan Hills—and the year it was found—1984—as well as the relative importance of the meteorite—1). ALH84001 made a huge impact in the science in 1996 when it was announced that the meteorite may have evidence for Martian bacteria in it. This argument was based on three criteria: the presence of microbial shapes from microscopy, organic compounds present within the meteorite, and the presence of small magnetite crystals.

The first two points were dismissed as evidence a few years later. Microbial forms are formed because of the preparation process used to examine the meteorite by electron microscope, and can be formed anywhere. Organics are also formed spontaneously when carbonate minerals are irradiated by lasers. The third point, the presence of magnetite crystals in the meteorite, is the only point not yet dismissed.

Magnetite, Fe_3O_4, is an important biomineral due to its magnetic properties. Microbes and multicellular organisms (including you!) form magnetite that serves as a natural compass. Knowing which way is north is useful for many organisms, as it removes one direction of uncertainty. Organisms form this magnetite with distinctive features. For one, biomineral magnetite is almost pure Fe_3O_4, whereas abiotic magnetite often has Ti and other contaminants. Additionally, the biomineral magnetite crystals are uniformly small, with sizes of about 20-30 nm, and these crystals have a specific crystal structure. The magnetite crystals within ALH84001 seemed to be biotic in origin, suggesting life.

Chapter 18 Practice Problems

1. Assuming the concentration of calcium does not change significantly in the oceans (0.01 M) and that the K_{SP} of calcite is 3.4×10^{-9}, what must the carbonate concentration be below the carbonate compensation depth?

2. If the concentration of Ca^{2+} in the Ordovician ocean is 400 ppm, and the concentration of Mg is 700 ppm, would this favor calcite or aragonite as a biomineral? Assume the TDS is 35000 mg/L.

3. A particularly nasty phosphate mineral is the mineral struvite, $MgNH_4PO_4 \times 6H_2O$, which forms kidney stones. This mineral forms naturally when physiologically dehydrated, where it can accumulate within the kidneys. The K_{SP} of struvite is $10^{-13.5}$. If your blood has a Mg content of 38 mg/L, and a phosphate content of 100 mg/L, what is the maximum NH_4^+ content that your blood can have before struvite starts to spontaneously precipitate?

Chapter 19. Deep-time Geochemistry

19.1 Origin of the Universe

This chapter will provide an overview of several big-picture science questions, united by the theme, "Where did it come from?" Several of these questions can be answered at least in part using geochemical techniques.

Most of the rock forming elements and elements that make up the Solar System originated in three main element-forming events. These events yielded the Solar System elements in the pattern we've discussed previously (Figure 19.1 or 3.2). These events include primordial nucleosynthesis, stellar nucleosynthesis, and supernova nucleosynthesis. Nucleosynthesis is the process by which new atomic nuclei are synthesized. Most of this occurs by the process known as *fusion*, which is the joining of two atomic nuclei to make a new atomic nucleus with greater mass. This differs from *fission* as fission goes the opposite direction, with smaller resulting atomic nuclei from the breakup of a larger atomic nucleus.

Figure 19.1. Cosmic abundance of the elements from Anders and Grevasse (1989)

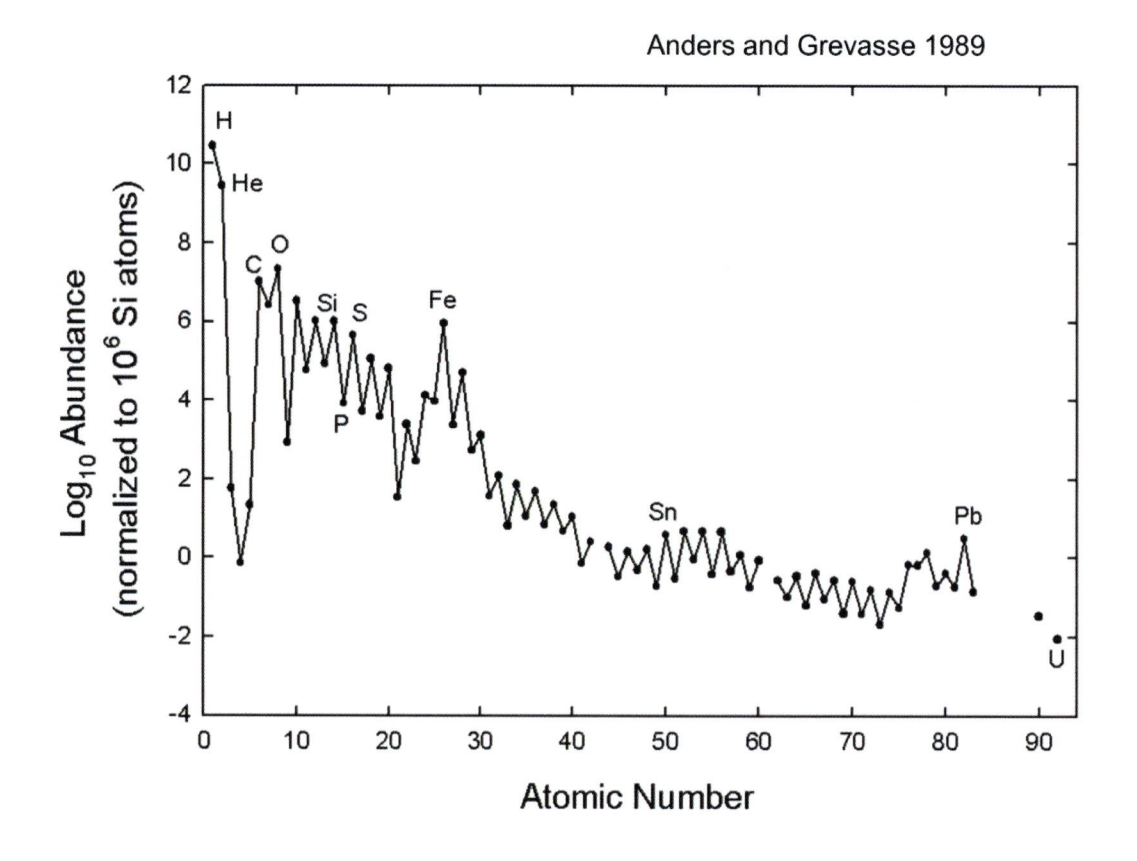

The elements occur with a saw-tooth pattern that results from their synthesis pathways. *Primordial nucleosynthesis* is the nucleosynthesis that occurred during the Big Bang some 13 billion years ago. In the first few minutes of the universe's existence, matter was dense yet cool enough to self-assemble into hydrogen atoms. These hydrogen atoms in turn fused to give helium and possibly lithium, but the expansion of the universe prevented further nucleosynthesis. For this reason, hydrogen and helium are the most abundant elements.

The universe was fairly limited in composition for a few thousands of years until material could start to accumulate into stars. The formation of the first stars allowed for *stellar nucleosynthesis* to begin, and resulted in the fusion of hydrogen into heavier elements, including helium and carbon. These elements in turn could fuse in the biggest stars, giving rise to all the elements up to iron. Iron is the last element that can form through stellar nucleosynthesis as fusion of elements heavier than iron requires energy, instead of yielding energy. This gave rise to the higher abundance of the elements carbon through iron.

Large stars capable of fusing elements to iron will also explode at the end of their life cycle, a process called a supernova. In doing so, they cast off large amounts of high energy neutrons and protons and other elements, which hurtle into nuclei, creating new, heavier nuclei. This process is *supernova nucleosynthesis* and results in all the elements heavier than iron.

The abundances shown in figure 19.1 are the abundances for the elements in the Solar System. As far as we know, the elemental abundance patterns in the bulk universe are probably quite similar. Some subtle changes have been observed in different stellar environments, but this pattern stands up as an average within an order of magnitude for most elements. One important consequence of these abundances is that the carbon to oxygen (C/O) ratio is less than one in our Solar System (and in most others). This ratio determines the types of rocks and minerals that can form rocky planets. Oxygen-rich systems will have silicate rock familiar to us, whereas carbon-rich systems will have carbide minerals, graphite, and diamond as major constituents instead.

19.2 Origin of the Earth

The Solar System has an age of about 4.56 billion years. This is younger than the 13 billion years of the universe. The origin of the earth and the Solar System has come about through thorough study of both meteorites and of star- and planet-forming regions in space. From astronomical observations, we have seen regions where dust and gas collapse by gravity, forming new stars (and probably planets). These regions may come about through a supernova shockwave propagating through a dense cloud.

Meteorites provide some evidence for a supernova. Some meteorites have an isotopic signature for ^{26}Al, a short-lived radioactive isotope of aluminum ($t_{1/2} \sim 7 \times 10^5$ years). For this isotope to have been trapped in rock, it would have to have been synthesized near the region where the Solar System formed, and near a supernova.

Meteorites also provide the clearest picture of how planets, such as the earth, form. The earth was formed from a mixture of meteorites. Meteorites bear specific oxygen isotopes (e.g., chapter 9) that show the earth originated from materials like these. Meteorites also provide a picture as to what happened after the meteorites accumulated into large bodies called planetesimals (smaller than planets, but large enough to be gravitationally bound together). Some of these planetesimals eventually became the asteroids, and some joined together to make the planets. Several asteroids appear to have differentiated rapidly, separating into a mantle and core within 10 million years of formation. The earth may have taken about that long as well.

19.3 Origin of the Moon

The Moon is an important body in the local space environment. The Moon provides tides, and may thus have allowed life to finally emerge onto land. The origin of the Moon is quite well known, having been solved because of thorough geochemical analysis of Apollo samples, coupled to efficient computational models.

Four primary hypotheses on the origin of the moon were suggested when the debate first came up. These were the 1) impact origin, 2) co-accretion origin, 3) spinoff origin, and 4) capture origin. These are shown schematically as Figure 19.2. In the capture origin, a small spherical planetesimal got caught in the earth's orbit, becoming the moon. The spin-off origin for the Moon suggested that the earth was spinning so rapidly early in its history that a chunk of the earth's mantle flew up and left the earth, melting to form the spherical moon. In the co-accretion model, the earth and Moon formed together from the same accreting cloud of dust and gas. In the impact model, the earth was smacked by a Mars-sized planetesimal, causing the earth to slosh out a blob of silicate rock that accreted to form the Moon.

Figure 19.2. The four hypotheses for the origin of the Moon. Earth is the larger blue-orange circle, with orange representing a metallic core.

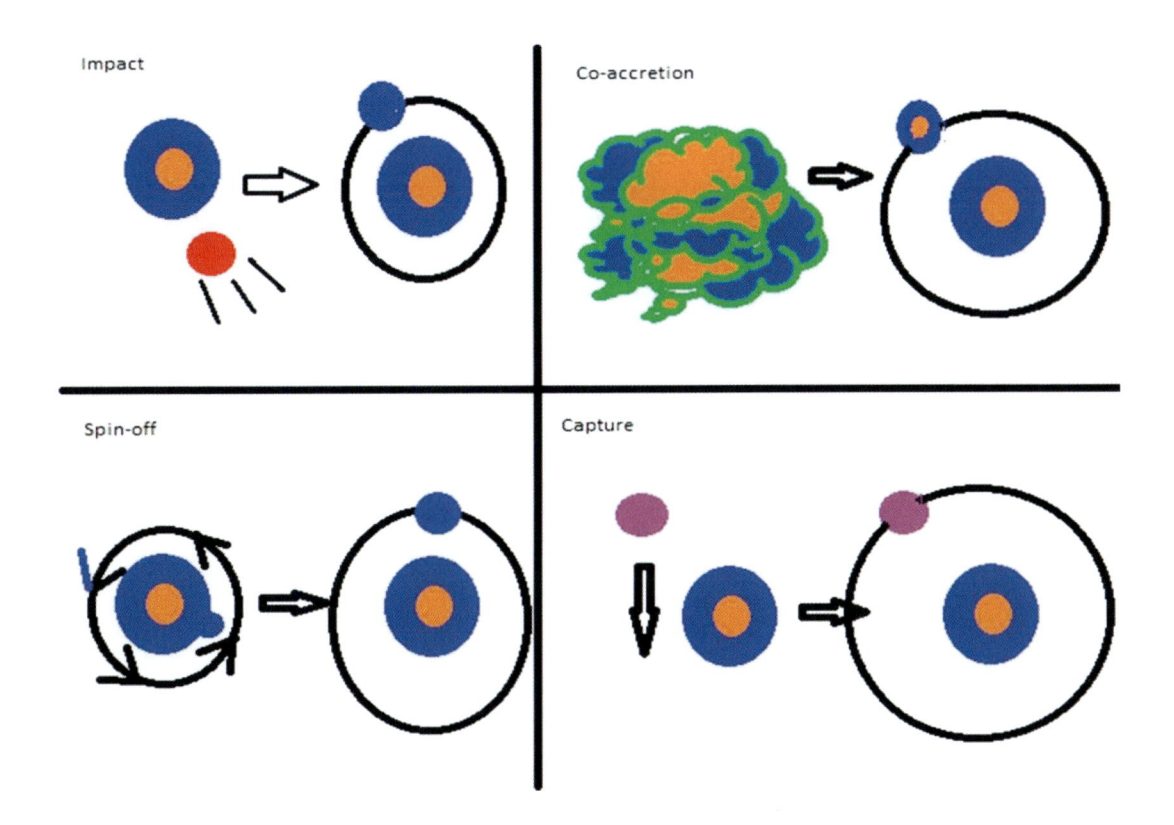

The Apollo moon missions revealed several interesting things: for one, the Moon had the same isotopic composition as the earth. This implied that the earth and the moon formed from similar material, in a similar region in space (eliminating the capture hypothesis). Additionally, the Moon does not have a large core, and is compositionally like the earth's mantle. This eliminated the co-accretion hypothesis. Finally, the physics required to spin off a chunk of rock to form the Moon require the earth to be going so fast that it probably wouldn't have stopped with just spinning off the Moon, eliminating the spin-off model. The impact model falls best with the data, as the Moon is compositionally like the earth's mantle and probably came from mantle material, but it doesn't have a core. It is believed that the Moon formed 4.4 billion years ago from the earth. If it formed by impact (and it probably did), then this event also likely liquefied the earth's surface, and would have boiled off any primordial oceans.

19.4 Origin of the Ocean and Plate Tectonics

So, when then did our oceans come about and how do we know? Similarly, when did plate tectonics begin? The origin of the earth's water has been the subject of some debate. Water is a volatile molecule, and given that the earth formed quite close to the Sun, relatively speaking, it should not have caught very much water during its formation. However, water, as stated previously, is common in the universe, and hence it could have been delivered by extraterrestrial material subsequent to the cooling of the earth (cometary delivery hypothesis). Alternatively, water trapped within hydrous minerals near the earth's surface could release this water on heating, steaming up an atmosphere that eventually rained down as water (the late veneer hypothesis).

The best data constraining the origin of the earth's water is the D/H ratio. Thus far, we've analyzed the D/H ratio of about three comets using spectroscopy, including Halley's comet, and found that the D/H ratio was significantly higher than the D/H ratio of the oceans. Alternatively, water-rich meteorites found in the outer reaches of the asteroid belt have D/H ratios much closer to the earth's oceans. The origin of the earth's water is still an open area of debate.

The origin of earth's oceans has been dated, however. Some of the oldest rocks on the earth can be found in Australia, and these old rocks include very old metasedimentary rocks. The location of major interest is at the Jack Hills range in Australia. Within these old metasedimentary rocks are sedimentary zircons that date to over 4 billion years in age. Zircons typically form in granitic rocks, which in turn form primarily in continental crust and through the action of water. As a result, these old zircons suggest both plate tectonics, and a global ocean were present as early as 4.4 billion years ago.

19.5 Origin of Life

One of the biggest questions that humans have asked is where life came from. Early ideas, continuing to this day, invoke supernatural sources, or wink*wink*nudge*nudge aliens (psst- it's still God!). The idea of spontaneous generation sought to address the origins question by arguing that living creatures spontaneously arose from non-living things. For instance, maggots spontaneously generate from rotting meat. This was dismissed for macroscopic organisms by a series of experiments by Franscesco Redi who showed that flies don't spontaneously pop out of rotting meat but instead are born from other flies. At the time, it was still believed that microscopic organisms could still arise spontaneously, though Louis Pasteur subsequently disproved that idea with careful experiments in the 19th century.

The scientific attempt to approach the origin of life began primarily with the work of Stanley Miller, who worked as a grad student to Harold Urey, a Nobel laureate geochemist who discovered the stable isotope deuterium. Miller set up a chamber with water being heated and stirred, in a chamber filled with methane and nitrogen. In this atmosphere, a spark discharged into the gaseous mixture. This process generated a tar that was condensed by cooling and reacted with the water. The water or other organics were analyzed. The material formed by this process included amino acids, simple carboxylic acids, and other biologic compounds. This experiment proved that biologically relevant organic molecules could come about from simple gases and energy.

Figure 19.3 The Miller-Urey Apparatus. The black stoppers had metal spikes inserted that propagated a spark, above a pot of water that was stirred. Note that this is a more modern approach to the apparatus, which originally involved multiple chambers, but this simple approach works just as well.

Subsequent work in this field has strengthened the original idea presented by this experiment, that organic compounds relevant to life could be formed easily by non-biological processes. Prebiotic material may have been abundant on the early earth. This led to an idea, probably wrong, that the early oceans were rich soups of prebiotic material.

The formation of biomolecules is an important first step in the origin of life. These biomolecules likely began as *monomers*, or single small molecules with 1-10 carbon atoms, a few nitrogens, some hydrogen and oxygen, and maybe phosphorus and sulfur. A possible early monomer might be adenosine triphosphate, or ATP (Figure 19.4). The next step, hypothetically, is to have these monomers assemble into polymers, which probably consisted of only one type (e.g., only amino acids or things similar, or only nucleotides). These polymers should have some chemical function, such as acting as enzymes. Ideally this chemical function is *autocatalysis*. Autocatalysis is the process by which a molecule (or a suite of molecules) makes a copy of itself from simpler monomers.

Figure 19.4. ATP monomer. While difficult to synthesize by non-biologic processes, once formed, this molecule could be part of the assembly material of RNA.

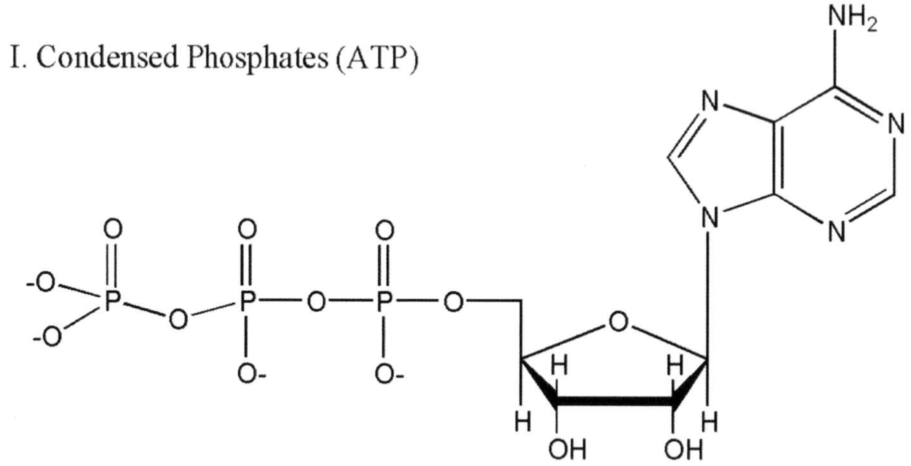

If these polymers have some chemical function or if they're autocatalytic, then chemical evolution can take place, selecting the most efficient molecules and increasing their number. Eventually a suite of polymers might have been present, with enough life-like features to be called life. We know that prior to the development of modern biochemistry, in which DNA codes for RNA codes for proteins, RNA was likely the sole informational molecule and possibly the main enzyme around. This time is termed the "RNA world" and likely preceded modern biochemistry.

Life likely arose sometime between 3.5-4.5 billion years ago. We do not have enough rock samples from this time to fully nail down the date, but isotopic evidence suggests life by about 3.8 billion years ago, and fossil evidence in the form of microfossils and stromatolites suggests life was present by 3.5 billion years ago. Life was clearly present by about 2.5 billion years ago, as it had begun to fundamentally change the chemical composition of the earth's surface.

19.6 Geochemistry of the Archean

As geologists well know, geologic time is divided up according to specific fauna present as fossils. However the time called "Precambrian" actually makes up 85% of the history of the earth, and is itself divided into several major eons. The Paleozoic, Mesozoic, and Cenozoic eras together make the Phanerozoic eon. Before the Phanerozoic eon is the Proterozoic eon, from 2.5 billion years ago to about 545 million years ago, and is the longest eon in earth history. Prior to the Proterozoic is the Archean eon, which spanned from 4 billion years ago to 2.5 billion years ago. Prior to that was the Hadean eon from 4.5 billion years ago to 4 billion years ago.

This chapter has covered most of the major events in early earth history. These events set the trajectory of the world on the path it is currently on, giving the splendor of life we see today. However, the Archean earth was quite different from the modern-day earth. For one, the earth's atmosphere varied significantly from present. The air you breathe today comes from a disequilibrium system generated by oxygenic photosynthesis. The atmosphere, prior to photosynthesis, did not contain very much O_2. In fact, the mineralogy of the Archean was quite different than present. Minerals such as iron pyrite (FeS_2) and uraninite (UO_2) were stable at the earth's surface, and in fact rolled around and were rounded by water. At present these minerals oxidize before they can be abraded and rounded to make pebbles.

The rise of oxygen from photosynthesis caused the formation of massive amount of iron deposits known as banded iron formations. These banded iron formations showed varying ratios of Fe^{2+} to Fe^{3+} minerals in different bands. The Fe^{2+} minerals attract a magnet and imply a lower redox state, whereas the Fe^{3+} minerals are not and imply an oxygenated environment. These rocks likely formed due to microbial photosynthesis within oceans that straddled the Fe^{2+}/Fe^{3+} redox boundary.

Chapter 19 Practice Problems

1. ^{26}Al has been detected in some of the oldest of meteorites. Given that its half-life is so short (7 x 10^5 years), how might this isotope be detected?

2. A Jack Hills zircon is found to have a ^{238}U to ^{206}Pb ratio of 1.1 to 1. Assuming there was no lead in the initial zircon, how old is this zircon? The half-life of ^{238}U is 4.47 x 10^9 years.

3. At what content of dissolved oxygen would you expect to find equal amounts of Fe^{2+} and Fe^{3+}, like those found in banded iron formations? Assume a pH of 7.

Chapter 20. Planetary Science: Water in Space

20.1 Water in Space

As we have mentioned before, water is one of the simplest molecules in the universe. It is the combination of the most common element in the universe, hydrogen, with the third most common element in the universe, oxygen. From this simple argument, we should expect to find water throughout the universe. Indeed, water is one of the primary building blocks of planets and bodies in outer space. Several figures showing evidence of water are provided in this chapter; these images come from NASA's photojournal website.

We must differentiate between water and liquid water. Ice and steam (steam may be more broadly referred to as water vapor) have properties quite different from liquid water. At a suitably cold temperature (100-200 K) water ice becomes as hard as the average terrestrial rock. Water also will differentiate during planet and moon formation, usually floating on top of rock and metal, where it forms a solid ice layer analogous to the mantle. Water as both an ice and a vapor make up a significant part of comets (Figure 20.1), as the sublimation of ice leads to violent ejection of water vapor, resulting in the distinctive coma of comets.

Figure 20.1. The nucleus of comet Wild 2, which consists of silicates and ices. From stardust.jpl.nasa.gov.

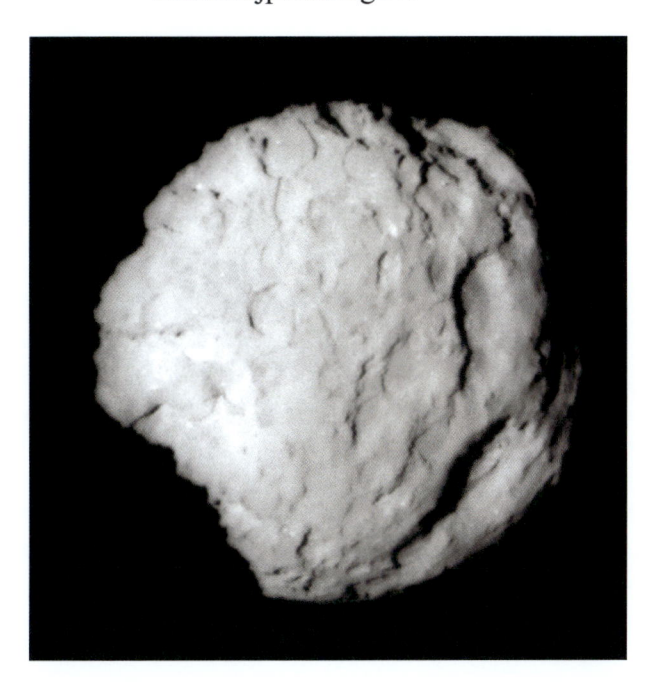

20.2 Water Phase Diagram

The phase of water (solid/liquid/gas) present at a specific condition is determined from the water phase diagram (Figure 20.2). This diagram shows the conditions for stability of a variety of water phases as a function of pressure and temperature. Two notable points are the triple point for water, which occurs at the point where all three phases of water coexist (around about 0.01°C and 0.006 atmospheres), as well as the critical point, the point where liquid water is no longer stable (218 atmospheres and 374°C). Additionally, the stability line for water at one atmosphere is shown, which shows that water is stable from 0 to 100 °C at 1 atmosphere. An important takeaway message from Figure 20.2 is that the stability field for liquid water is the most limited of the three water phases, and that liquid water requires decent pressure and moderate temperatures.

Figure 20.2 Water P-T phase diagram. Not to scale. The stability region of liquid water is the smallest, as vapor extends beyond the triple point (374°C and 218 atm), as does ice.

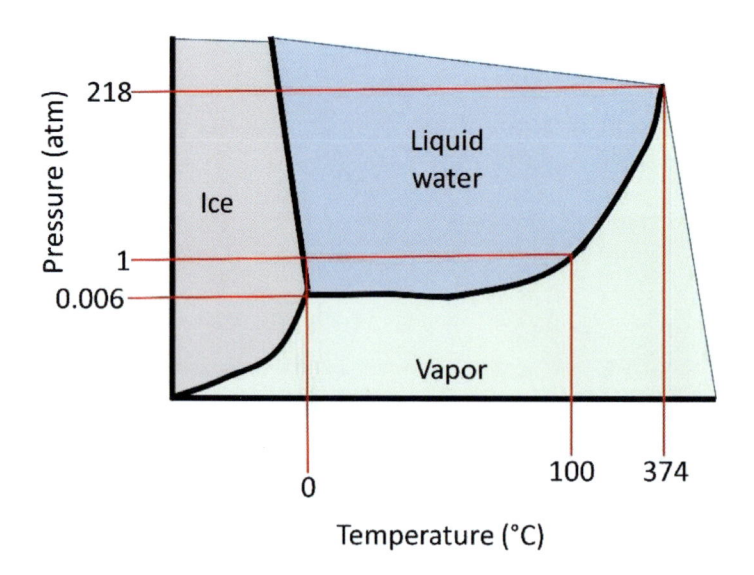

Water changing from solid to liquid melts, and the reverse is freezing. Water changing from liquid to gas boils, and the reverse is condensing. Water changing from solid to liquid sublimates, and the reverse is deposition. The physical parameters of each of these processes are outlined in chapter 3.

Detection of water, especially of liquid water, is a major goal of planetary science. Liquid water is presumably necessary for habitability, and detection of liquid water, either directly or indirectly, in turn identifies extraterrestrial habitats where life might live.

20.3 Water Detection

The detection of water takes place through a few primary means: we can detect the mineralogical evidence of it in samples from other space bodies, we can determine its presence spectroscopically, and we can indirectly detect it through other physical means. Sometimes we can see it directly, or see its evidence from changes to planetary morphology. Many of these methods have been used to search specifically for liquid water, as opposed to ice or water vapor, both of which are much more common in the universe.

Mineralogical evidence: Several minerals form only in the presence of water, either as liquid or vapor. These include many minerals identified in the metasomatism chapter (chapter 10), include clays and other phyllosilicates, amphiboles, and some carbonates. We have several meteorite samples that have significant quantities of clay minerals, as well as some Martian meteorites with carbonates that appear to have been deposited as the result of water on Mars.

Figure 20.3. Hydrogen abundance measured at the asteroid Vesta's surface by the Dawn spacecraft. Hydrogen is probably correlated to water ice or to hydrous minerals (see https://www.nasa.gov/mission_pages/dawn/multimedia/pia16180.html).

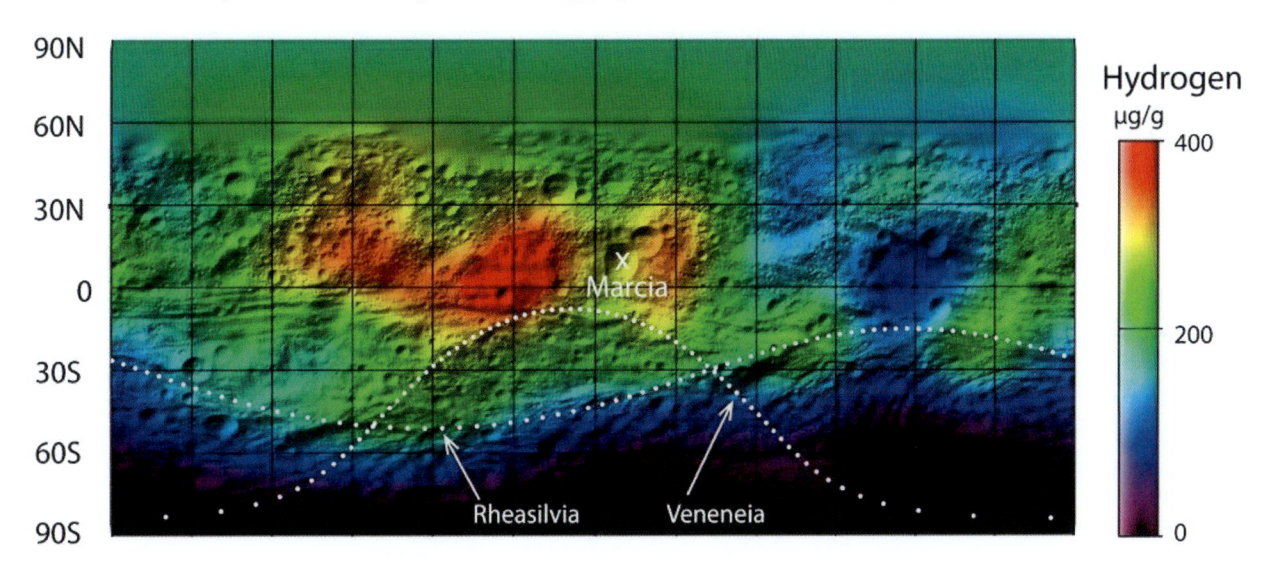

Spectroscopic evidence: Water has distinct spectroscopic signatures, and the identification of these signatures points to liquid water. Comets have spectra with distinctive features from water vapor, and it is these features that allow us to call comets "dirty snowballs". Martian water ice has been detected from the neutron flux of the Martian surface, since neutrons are readily absorbed by water (or rather, by the protons in water), Martian environments rich in water ice have a low neutron flux. Water on the asteroid Vesta has been detected using tools similar to these (Figure

20.3). Additionally, water in the plumes of Saturn's moon Enceladus has been detected using both IR spectroscopy and mass spectrometry (Figure 20.4).

Figure 20.4. Enceladus, a moon of Saturn, has a geologically active region in its south that has deep grooves called "Tiger stripes" (top panel- see https://www.jpl.nasa.gov/edu/news/2015/10/23/flying-by-saturns-moon-enceladus/). Water vapor jets out of these stripes, as measured by mass spectrometry (bottom panel- see https://saturn.jpl.nasa.gov/resources/4015/).

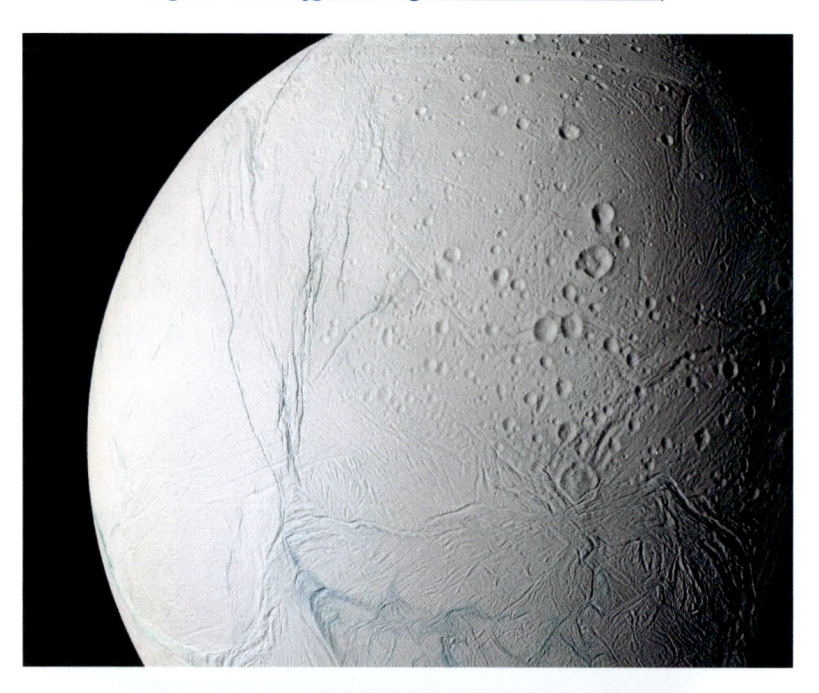

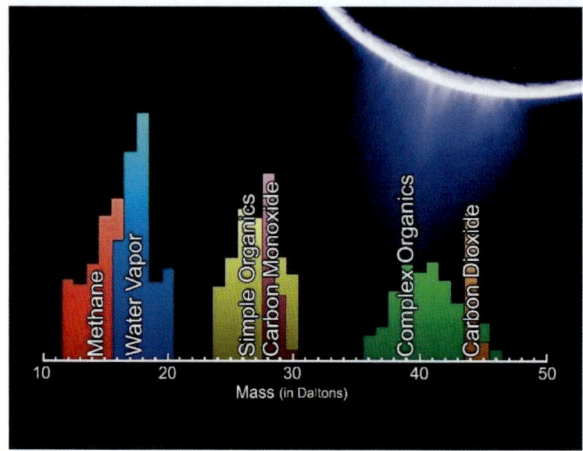

Geophysical evidence: One way to identify water indirectly is to hunt for geophysical evidence for water on a planet. This can include trying to find a conductive signature (since water rich in salts is highly conductive), which is indicative of a large liquid body (this is how the subsurface ocean of Jupiter's moon Europa was identified).

Alternatively, since liquid water is a fluid, it is tolerant of bending and flexing, and hence liquid water may be detected on moons that are subjected to strong tidal stressing by investigating the flexure of these moons.

<u>Planetary morphology evidence</u>: Some of the strongest lines of evidence for water, especially of liquid water come from planetary morphology. Mars is the best example of this. From a distance, we have observed gullies, fluvial deposits covering other rock types, mud cracks, and even river valleys on the Martian surface (Figure 20.5). Additionally, water ice is common on Mars' surface, and is commonly found at the bottom of craters. Jupiter's moon Europa also has a surface that is best characterized by the moving of water-ice plates on a liquid water ocean, which make for a clear physical geology exercise (Figure 20.6).

Figure 20.5. View of the Martian surface with gullies, possibly but not necessarily, carved by liquid water. See https://www.nasa.gov/feature/jpl/mars-gullies-likely-not-formed-by-liquid-water.

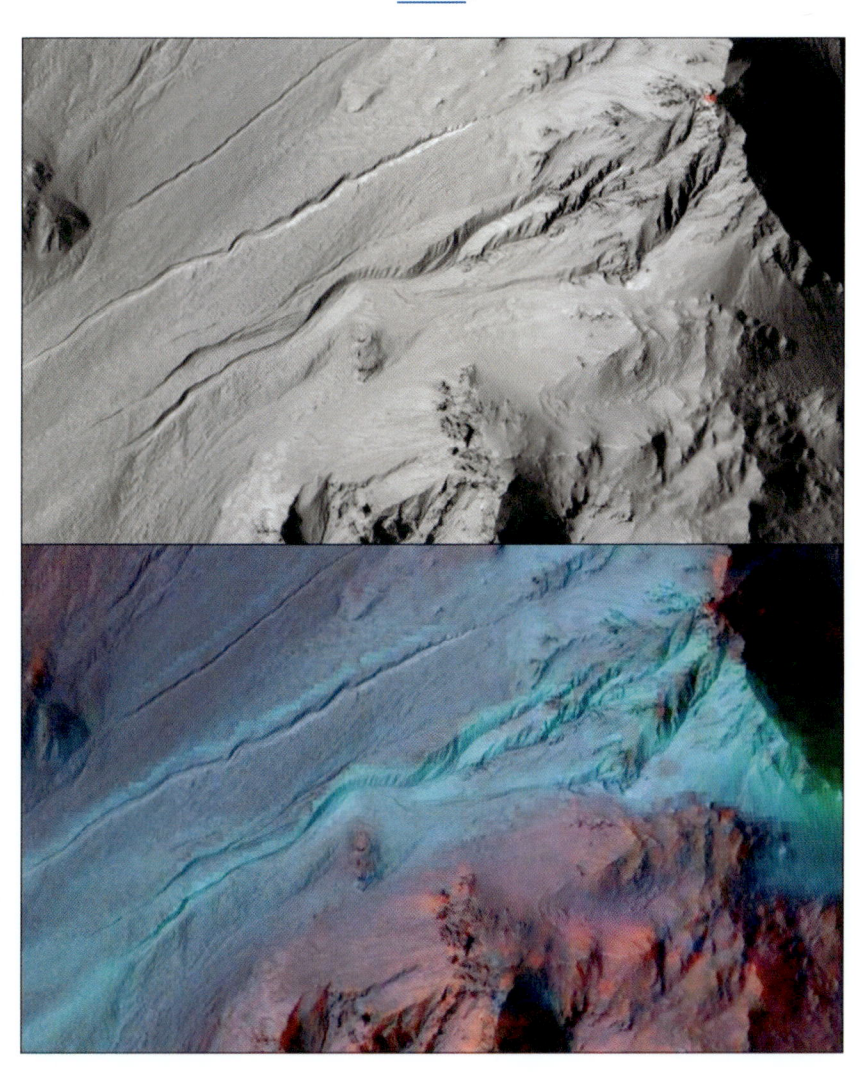

Figure 20.6. Europa's surface is characterized by plates that have moved relatively recently, with several moving events having occurred over a relatively short geologic time. From https://www.nasa.gov/multimedia/imagegallery/image_feature_529.html

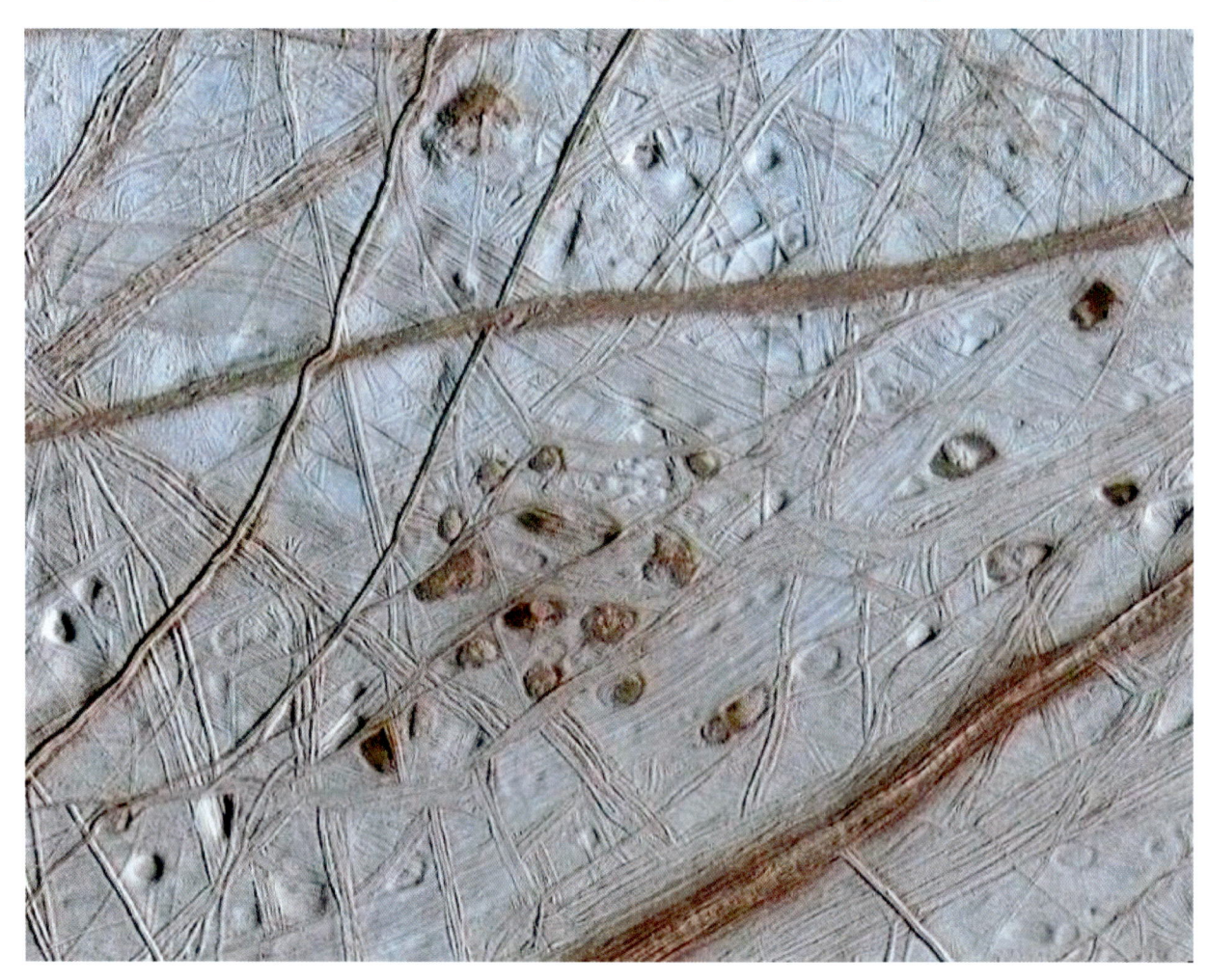

Chapter 20 Practice Problems

1. The surface pressure of Mars varies between 5 and 7 mbar, and the temperature ranges from really cold (-248 °F) up to a warmer 70°F in peak sun (sometimes even hotter). Can liquid exist on the surface? Why or why not?

2. The water coming out of Enceladus has been measured to have both CO_2 and organic compounds such as methane. If the pH of the Enceladus water is about 8, what is its redox potential (check out chapter 7).

3. Using the map of Europa below, determine the order in which the cracks A-D occurred.

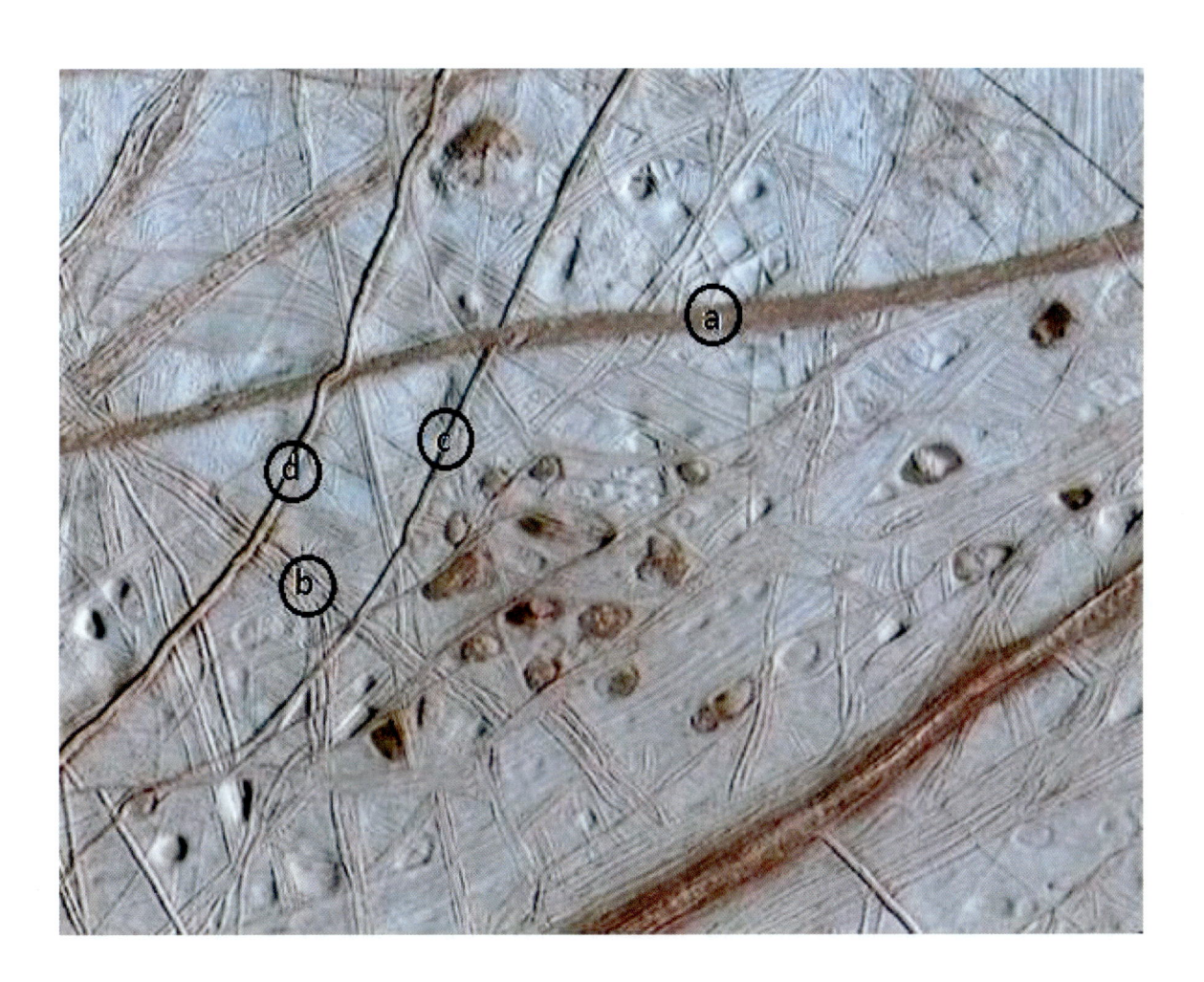

Chapter 21. Liquids on other Planetary Bodies

21.1 Titan

Much of the unique geology on earth, and to a lesser extent on Mars, comes about because of the action of liquid water at its surface. An open question for planetary scientists is that, since water seems to drive much of the geology of the earth, must liquid water be necessary for these unique processes or might other liquids also drive similar geomorphological processes on other worlds?

Saturn's moon Titan presents us with an excellent test case. Titan is one of the largest moons in the solar system; only Jupiter's moon Ganymede is larger. Titan is also the only moon in the solar system with a substantial atmosphere. This atmosphere consists of a mixture of N_2 and organics, mostly methane, but with a few others, too. The surface of the moon cannot be seen in visible light as the atmosphere is full of an organic-rich haze. The temperature of Titan's surface is about 90 K, and the surface pressure is about 1.4 atmospheres. As Titan is at Saturn, which is about 10 times further from the sun than the earth, it receives about $1/100^{th}$ of the sunlight.

21.2 Atmosphere and Methane

Much of Titan's uniqueness comes from its thick atmosphere. Compositionally, Titan consists of a rocky core covered with water ice. The surface rock of Titan is in fact ice (see the following chapter 22 to see more examples of this elsewhere in the Solar System). However, at the pressure and temperature of Titan's surface (~90 K), ice is as hard as rock, and no water can effectively turn into vapor.

To repeat, the temperature of Titan's surface is about 90 K, and the surface pressure is about 1.4 atmospheres. While this temperature is nowhere close to the earth's surface temperature, the pressure is the closest to the earth's pressure of all the other places in the solar system. Additionally, this atmosphere is 98% N_2, and the earth's is 78% N_2. The main difference between these two atmospheres' composition is the lack of any oxidizing compounds on Titan, such as O_2 or CO_2. Titan's surface is extremely reducing, with CH_4 comprising much of the remainder of the atmosphere.

The P-T diagram for methane is shown as Figure 21.1. Like water, under low pressure methane ice sublimates directly to gas, but as pressure increases, there is a zone of stability for liquid methane. If this is true, then might we expect to see methane (and other, similar hydrocarbons) acting to modify Titan's surface, like water does to the earth?

Figure 21.1. Schematic diagram of methane stability in P-T space. Data from airliquide.com

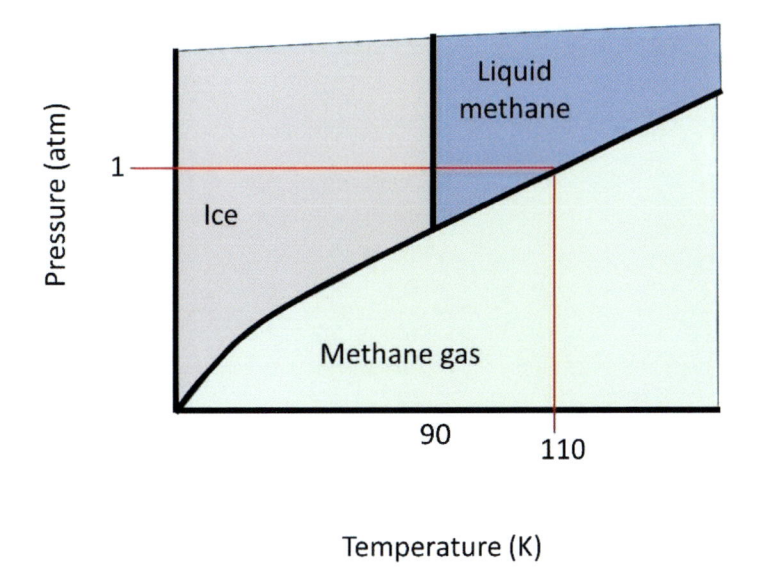

21.3 The hydrocarbon cycle on Titan, and the Cassini-Huygens spacecraft

The water cycle was discussed in chapter 3. It consists of the evaporation, respiration, and transpiration of water, condensation of water (in precipitation), surface and groundwater flow of water, and storage of water in the ocean. Due to its thick, hazy atmosphere, we cannot easily see the surface of Titan using telescopes. The Cassini-Huygens space craft has provided nearly all our knowledge of the surface of Titan. This spacecraft can view the surface of Titan using radar or by looking through certain spectroscopic windows in the infrared spectrum (Figure 21.2). Radar is not absorbed by the hazy hydrocarbons, and beamed down by the radar instrument on Cassini to Titan's surface. Smooth material comes back dark, whereas rough material comes back light. Finally, the Huygens probe entered the atmosphere of Titan, and took pictures down along its descent, including a few final pictures at the surface of the moon.

Figure 21.2. The Visual and Infrared Mapping Spectrometer (VIMS) instrument on Cassini generated this map of Titan by looking through spectral windows. Picture from https://saturn.jpl.nasa.gov/resources/6278/

If there is an active liquid cycle on Titan, we should expect to see precipitation, condensation, flow on the surface, and transport to large liquid reservoirs. Indeed, clouds are common at Titan's poles (Figure 21.3), and these clouds move with time. Indeed, clouds can be seen using the VIMS instrument at the south pole. Upon the observation of clouds on Titan by Cassini, researchers revisited Hubble data on Titan, and detected the presence of clouds in those images as well, though they were not recognized at the time.

Figure 21.3. Titan's clouds (white) evolve over time. Image from https://photojournal.jpl.nasa.gov/catalog/PIA06110

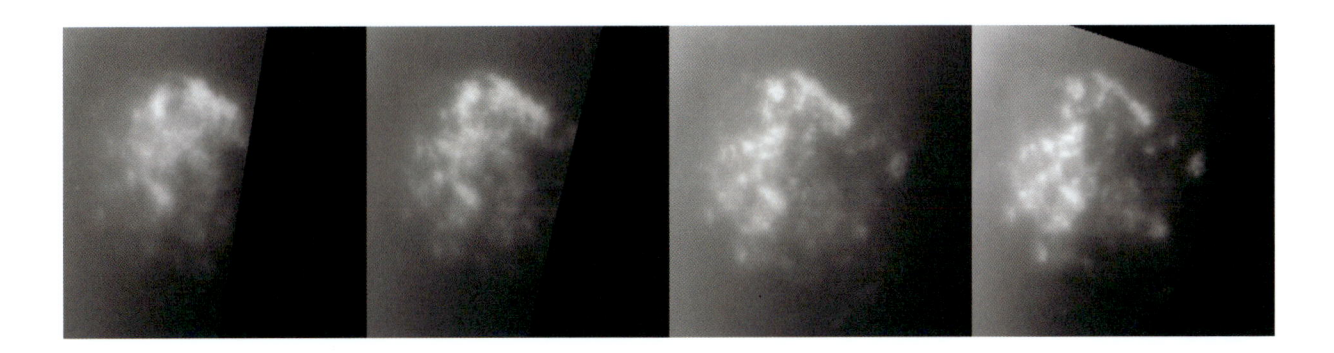

Clouds have also been seen elsewhere in the solar system, including on Mars, and across Venus. The gas giant planets are completely covered in clouds. Precipitation coming from the clouds would be much more indicative of a liquid cycle than the mere detection of clouds. Precipitation is difficult to detect, but flow features are not. Flow features include meandering rivers, streams, branching river patterns, and alluvial fans.

As the Huygens probe approached Titan's surface, the Descent Imager/Spectral Radiometer (DISR) instrument took pictures of the approaching ground. As it did so, it saw some intriguing features (Figure 21.4). These features include channels, and a branching pattern characteristic of rivers. These rivers emptied into a medium-gray region to the south east. The Huygens lander landed in this medium-gray region.

Figure 21.4. Branched valleys observed by DISR on the Huygens probe. Image from https://nssdc.gsfc.nasa.gov/planetary/titan_images.html

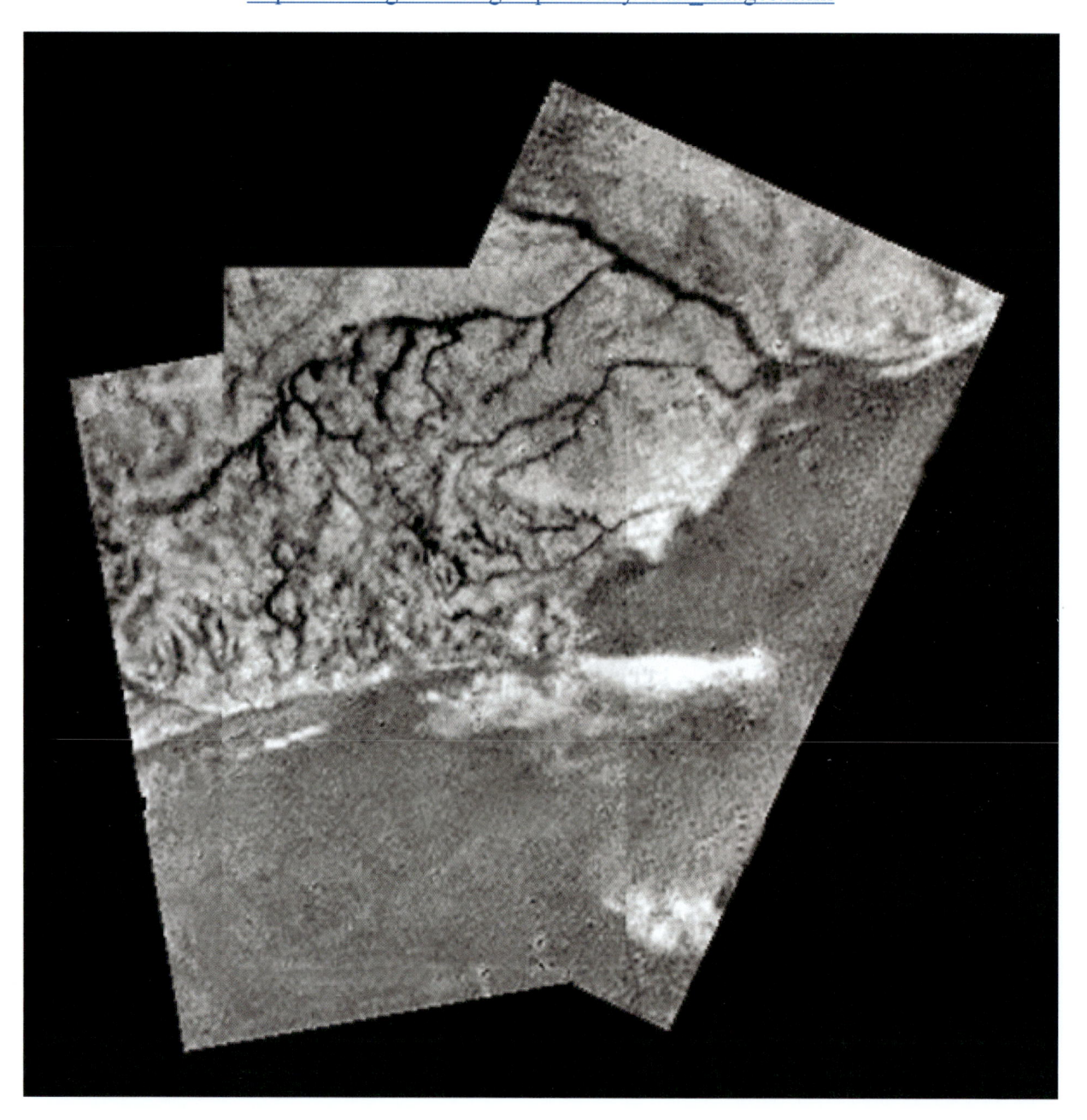

In addition to images obtained during descent through Titan's atmosphere, the radar maps of Titan also provided a valuable look at Titan's surface. Dunes are observed near the equator of Titan, as are some craters and possibly cryovolcanoes. With respect to a liquid cycle on Titan, the north pole of Titan proved to be the most fruitful (Figure 21.5). Lakes of liquid hydrocarbons are spotted across the north and south poles of Titan.

Figure 21.5. Radar image of lakes at Titan's north pole. Image from
https://www.nasa.gov/feature/jpl/experiments-show-titan-lakes-may-fizz-with-nitrogen

A key proof of a liquid cycle occurring on Titan was discovered when the Huygens probe crashed onto the surface. The cameras aboard this probe continued to take some pictures even after landing, and what these pictures revealed have not been observed anywhere else in the Solar System outside of the earth (Figure 21.6). Round rocks! Consider on Earth what round rocks imply. Smoothing a rock to roundness typically requires having the rock become abraded and transported down a river or stream. These rocks (which are likely made of ice) indicate that liquid flow is common to the surface of Titan. Recall also that the probe landed within the gray southeast region of Titan (Figure 26.4). This region is not a body of water, but was a solid surface. Although the exact texture of the surface is unclear, it either consisted of a pebbled soil, or a frost-covered soil, but was generally dry.

Figure 21.6. Image from the surface of Titan. The round rocks are about 10 cm across.
https://nssdc.gsfc.nasa.gov/planetary/titan_images.html

Titan is the only other planetary body we've seen where liquids are actively modifying its surface on a large scale. Water may be active on Mars and Europa, but rain, flow, and evaporation are best observed on Titan.

Chapter 21 Practice Problems

1. Let's say an unknown organic (MW 78 g/mol) has a solubility within liquid methane of 10 g/kg. What is its molarity? The density of liquid methane is 423 kg/m³.

2. If these organics are soluble in methane, but not volatile, what might happen to them as they are transported in the liquid hydrocarbon cycle? Is there an analogous process on earth?

3. The Huygens lander hit the ground going about 30 m/s. It has a mass of 319 kg. How much energy did it transfer to the ground? Assume that this energy went into vaporizing liquid methane, which has a heat of vaporization of 510 kJ/kg, how much methane did it vaporize? If the surface has a porosity of sand at 30%, what volume of methane around the probe did this evaporate?

Chapter 22. Water as Rock

22.1 Water and Moons

By now you should see a general theme to fluids in space. Water is common, but exists mostly as a solid (Chapter 20), and is often detected by its effects, both mineralogy and morphological. Other liquids, such as organic hydrocarbons, can substitute for water and can participate in a fluid cycle analogous to the water cycle (Chapter 21). In this chapter, we shall cover the icy bodies in the outer Solar System, discussing their features, and what they tell us about geological processes outside of the earth.

Figure 22.1. Diagram showing inclination (i) and prograde (p) and retrograde orbits (r).

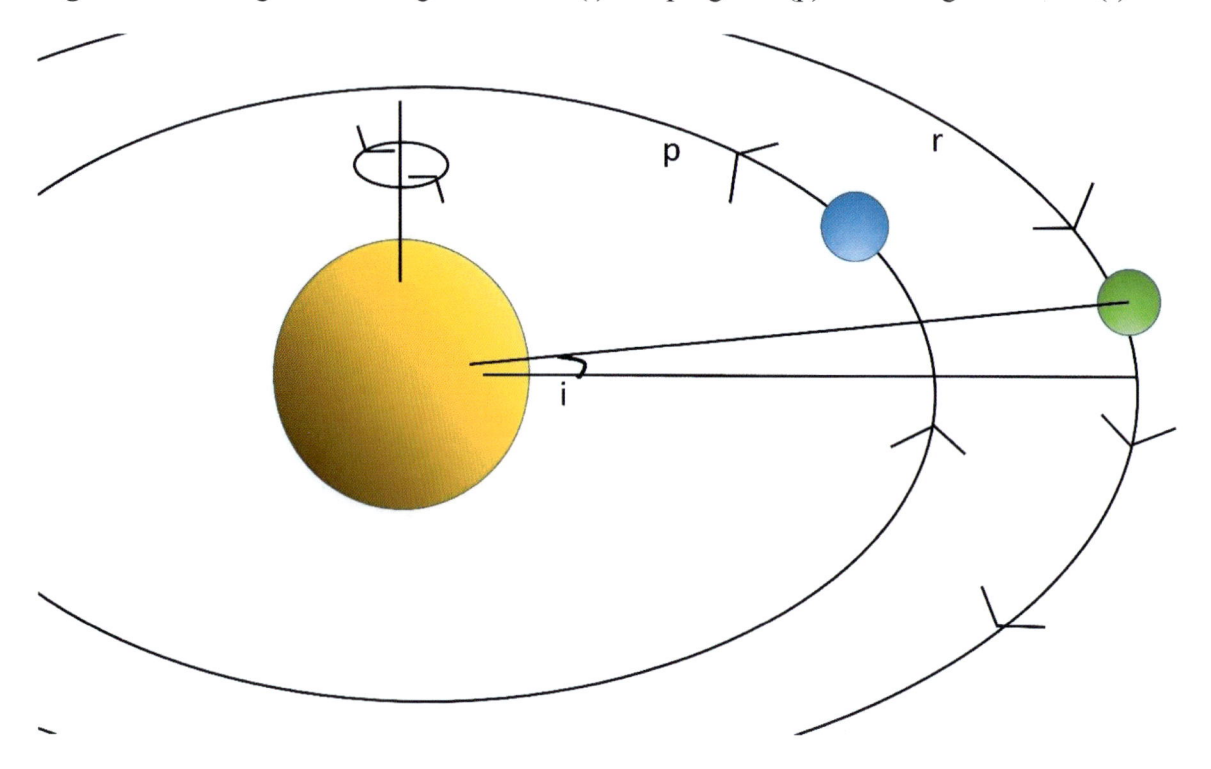

A moon is a small body that orbits a much larger body. In most cases the moon is significantly less massive than the body it orbits; an exception is the Pluto system (discussed below). The earth-moon system is close to also having this issue, but luckily the earth is significantly larger than the moon. A moon around a gas giant planet can be rather large: the largest moons in the solar system (Ganymede and Titan) are both larger than the smallest planet, Mercury. Large moons will be spherical because of their own gravity, but small moons may not be large enough to become spherical. Some moons may have a significant *inclination*, which is the difference between their angle of orbit and the plane of rotation of the planet around which they orbit (Figure 22.1, angle *i*). Many planets have *prograde* orbits, or they orbit in the same direction that the planet rotates. Those that orbit in opposite directions have *retrograde* orbits. A *regular* satellite is one that has a small inclination and is prograde. Regular satellites probably formed with the planet 4.5 billion years ago. An irregular satellite is one that has a large inclination and/or is retrograde. Most irregular satellites were probably captured in the gravitational field of the planet they now orbit.

22.2 Jupiter's Moons

Jupiter is the most massive planet in the Solar System. In some ways, it is like a failed star. It even formed its own mini-Solar System. Jupiter has many moons, but four dominate. These four are the *Galilean* satellites Io, Europa, Ganymede, and Callisto.

Io is the only moon we will discuss in this chapter that lacks water. It is the densest of Jupiter's moons, with an average density of about 3.6 g/cm^3. It surface is the youngest in the Solar System, and is covered with volcanoes and fresh basaltic lava fields. We have observed its surface for over 20 years, and in that period new volcanoes have erupted at its surface (Figure 22.2). This extreme volcanism is driven by tides from interaction with Jupiter and Europa.

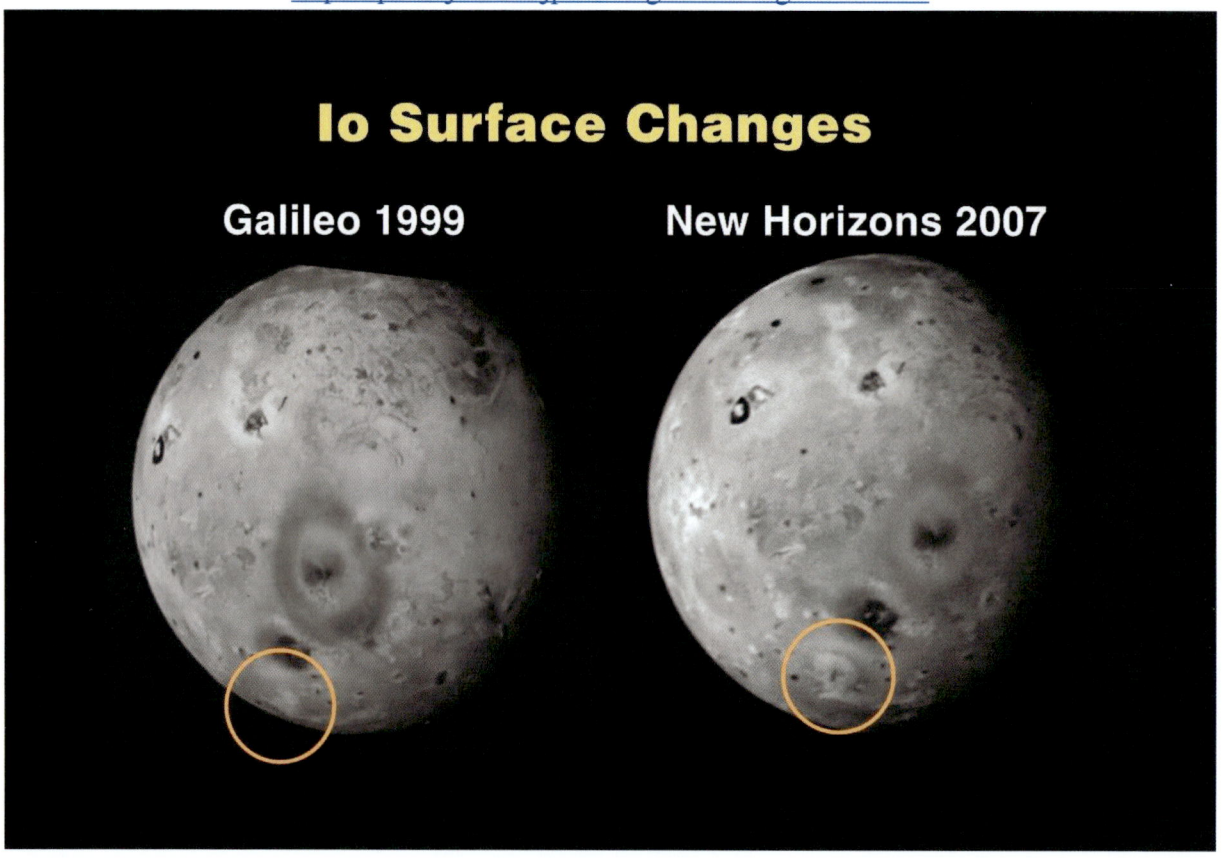

Figure 22.2. New volcanoes have erupted on Io over 8 years. From https://photojournal.jpl.nasa.gov/catalog/PIA09355

Europa is an ice-covered moon and has a young surface, as evidenced by the lack of abundant craters. The major geologic feature on Europa are grooves caused by flowing ice sheets (Figure 22.3). Beneath Europa's icy crust is likely a deep liquid water ocean that is in direct contact with a rocky mantle. It is the most spherical of all planetary bodies. Europa has a very thin oxygen atmosphere. This atmosphere comes from the high-energy particle (hν) bombardment of Europa's surface, which splits apart water molecules to make hydrogen peroxide, oxygen, and hydrogen:

$$2\ H_2O + h\nu \rightarrow H_2 + H_2O_2$$

$$2\ H_2O + h\nu \rightarrow 2\ H_2 + O_2$$

Figure 22.3. This false color image of Europa shows the grooved terrain. From
https://www.nasa.gov/multimedia/imagegallery/image_feature_1339.html

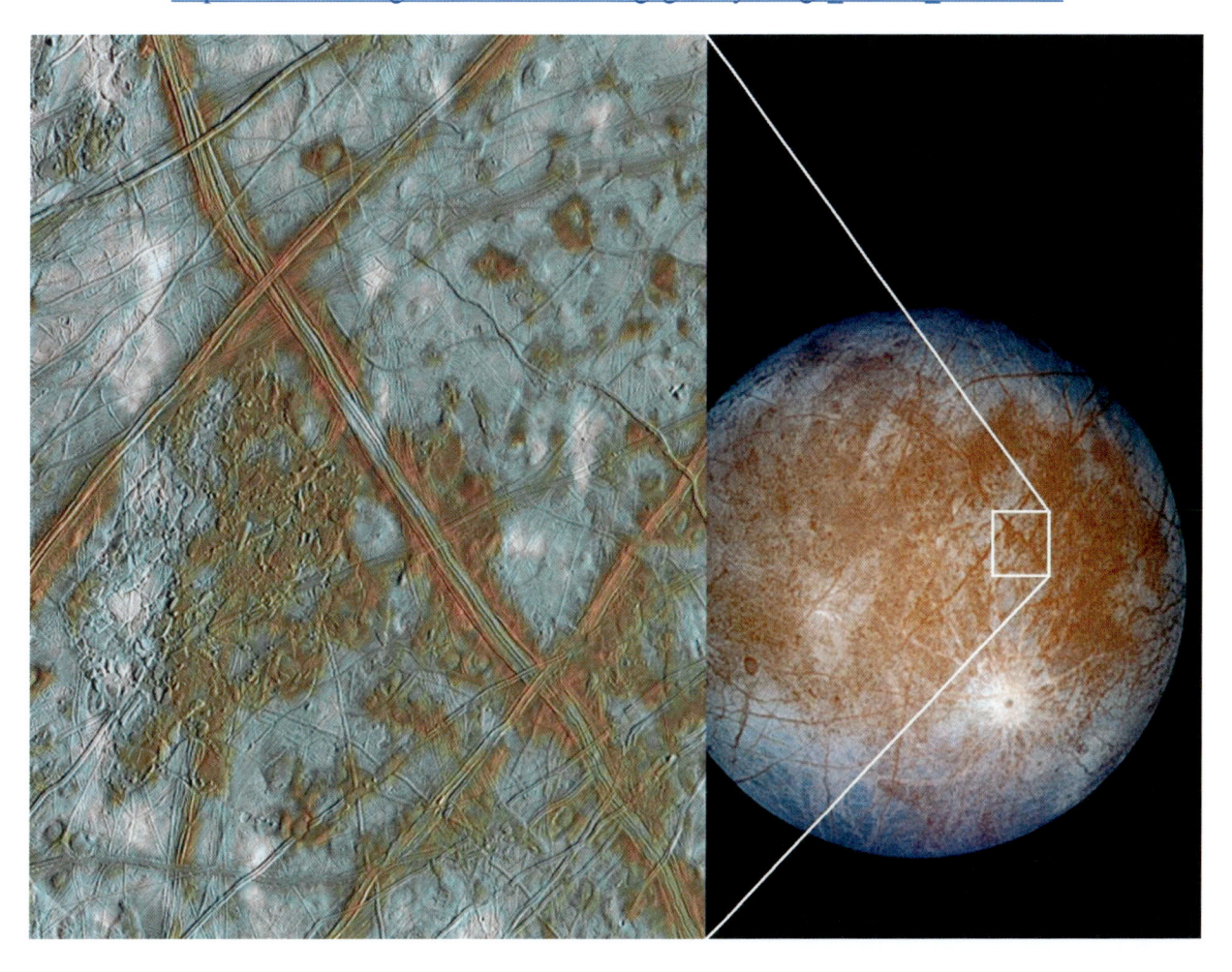

Ganymede is the largest moon in the Solar System. Its surface is mostly ice with some silicates, and is old. Some of the terrain looks grooved and cracked like Europa's surface, and other portions are only covered with craters (Figure 22.4). Like Europa, Ganymede has a thin oxygen atmosphere.

Callisto is the most distant of the Galilean satellites from Jupiter. Its surface shows relatively little geologic activity beyond cratering. The surface of Callisto is old, it is covered in craters (Figure 22.4), and consists of ice, silicates, and some volatile materials. Callisto appears to be undifferentiated, and did not separate into a crust, mantle, and core.

Figure 22.4. Ganymede (left) and Callisto (right) both have old surfaces covered with craters. From https://www.jpl.nasa.gov/spaceimages/details.php?id=PIA01299

22.3 Saturn's Moons

Saturn is smaller than Jupiter. It has several moons (Figure 22.5), of which Titan is the largest and most geologically unique (Chapter 21). Several of the smaller moons are spherical, and hence will be discussed briefly. Most of these moons are made of water ice with small amounts of rock. Except for Titan and Enceladus, most of these moons are geologically dead. The main features of these moons are craters. Some have grooved terrain like Europa and Ganymede, and some have some cliffs, but in general, most are cratered ice spheres with little else going on. The temperature here is about 100 K, so water ice is easily as hard as rock.

Other moons around Saturn include Mimas, which has a huge crater, the formation of which nearly destroyed the entire moon. It is the smallest round object in the solar system, and sure looks like the Death Star. Tethys is larger, and is made of almost pure water ice. It has a very bright surface. Dione and Rhea are denser than Tethys and both likely have silicate cores. Iapetus has a large equatorial ridge, probably formed by a large impact which almost destroyed the moon. It has a significant surface contrast with a bright white material covering part of the moon (likely deposited volatiles) and a dark material, probably organic goo.

Enceladus merits its own paragraph. Enceladus is not a large moon, but it has a geologically active south pole. Most of Enceladus is covered with an icy crust covered with craters, suggesting that it is geologically dead. However, at the south pole of Enceladus, you can find a series of features called "Tiger stripes". These grooved features are young, warm (180 K) and appear to be releasing copious amounts of water. This water in turn crystalizes to form a ring of water ice around Saturn (termed the E ring). The source of energy powering these geysers or cryovolcanoes is unknown.

Figure 22.5. Saturn's moons Mimas (top), Enceladus (middle), and Iapetus (bottom, showing two colored terrain). From https://solarsystem.nasa.gov/planets/saturn

22.4 Uranus and Neptune

We know the least about the moons of these two planets as we have only visited each planet once with the Voyager 2 spacecraft. It does not appear as though we are missing much with Uranus, as there are no huge moons. There are five spherical moons, but most are small and made of water ice. Neptune, on the other hand, has one large moon, Triton. It is the coldest large moon in the solar system, even colder than Pluto.

Miranda is the smallest spherical moon of Uranus, and has deep grooves on its surface, indicating geologic activity at some point in its past. Ariel also has a grooved surface, but is larger. Umbriel is about the same size as Ariel, but doesn't have a grooved surface (or at least the part we've seen doesn't), and its surface is darker than the other moons, likely due to organic carbon covering some part of the moon. Titania and Oberon are slightly brighter, but also appear geologically dead. Very little is known about these moons, and it is unlikely at present that anything more will be learned about these moons soon, as no missions to Uranus are planned or even in planning for the next few decades.

Figure 22.6. Uranus's moons Miranda (top) and Umbriel (bottom). More here: https://solarsystem.nasa.gov/planets/uranus/moons

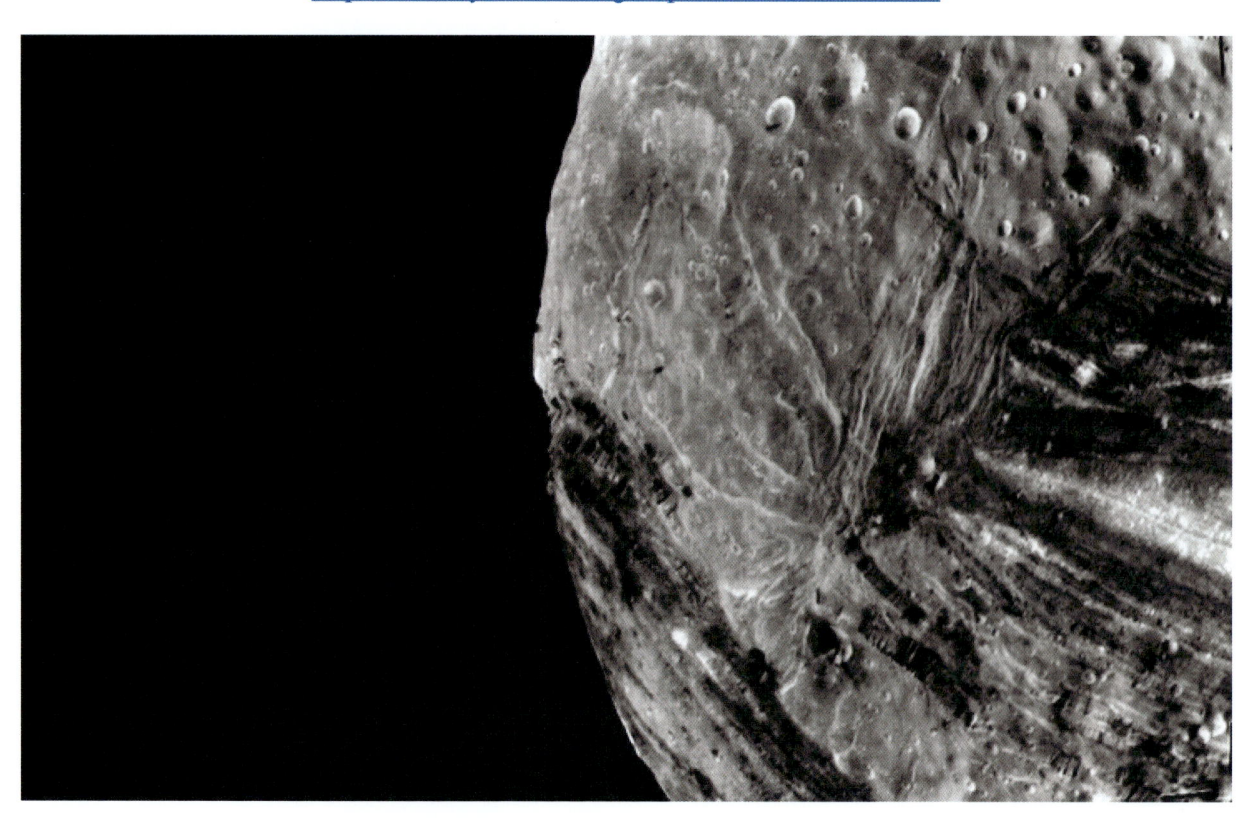

Triton is the only large, retrograde moon. It is likely a captured Kuiper Belt object, and is larger than Pluto. Triton has thin atmosphere consisting of N_2, which also deposits as an ice at the surface. The moon is geologically active, but the process that is driving this geologic activity is unknown. There are no terrestrial analogs to the surface of Triton, as the surface is colder (40 K) than anywhere on the earth. The surface is young, with relatively few craters. Triton's surface is broadly divisible into two terrains: the cantaloupe terrain, and a mottled white terrain (moldy cantaloupe?). These terrains may have originated from cryovolanism, or through the rising of cold, light material to the surface, a process called diapirism. Like the moons of Uranus, very little is known about Triton.

Figure 22.7. Neptune's moon Triton.

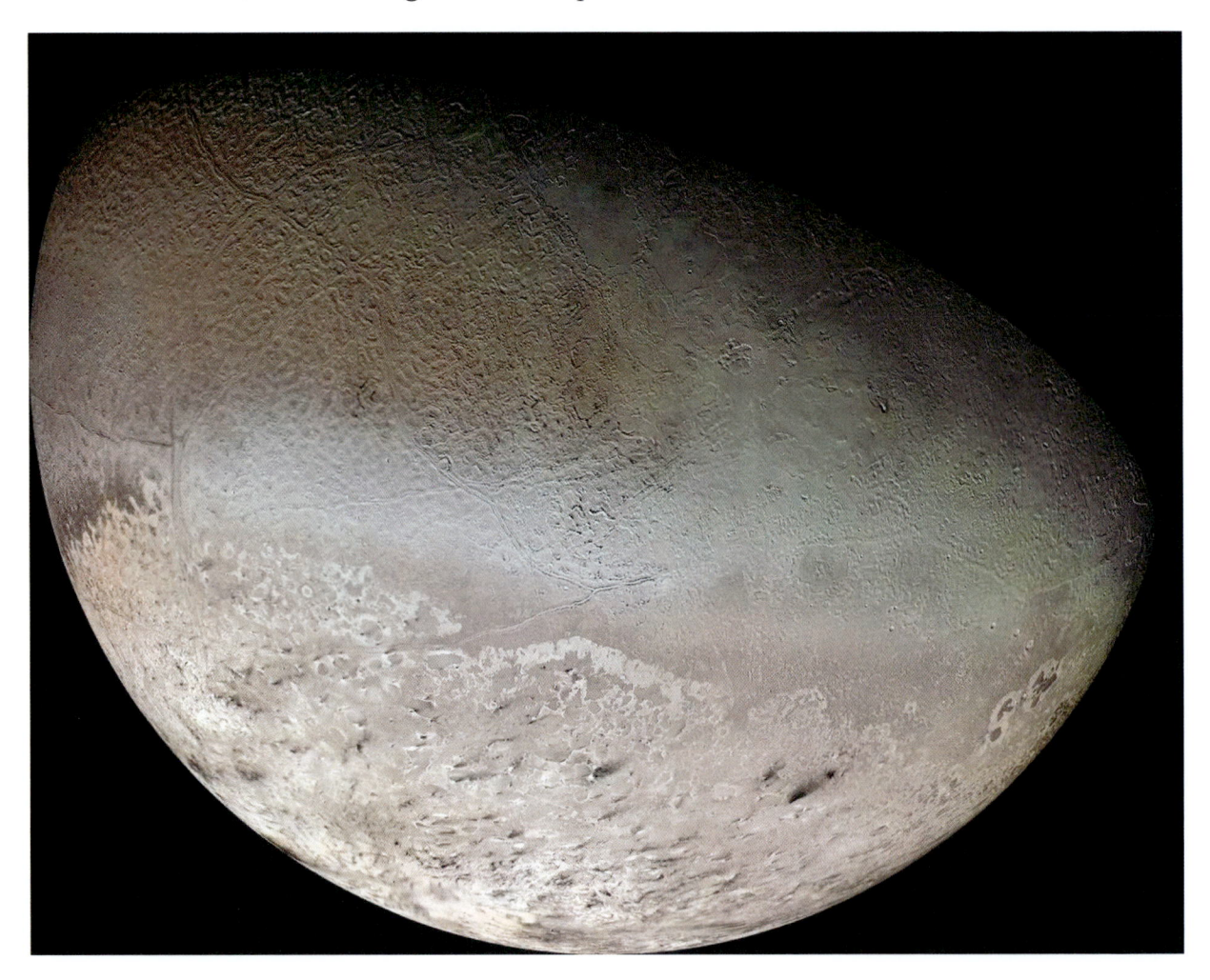

22.5 The Kuiper Belt and the Oort Cloud

The prior sections have discussed several intriguing moons in the solar system. Many of these are larger than Pluto. Pluto was discovered in 1930 by Clyde Tombaugh, and holds a special place in the US as being the American Planet™. However, Pluto has several features that make it unusual when compared to other planets in the Solar System. For one, it has a lot of moons (5 at last count). The largest, Charon, is big enough that the center of mass of the Charon-Pluto system is not within Pluto (all other planet-moon systems have their centers of mass within the planets). Additionally, it has a highly elliptical orbit, coming as close as 29 AU from the sun, bringing it within the orbit of Neptune, and going as far as 49 AU.

The official definition of a planet includes the following 1) it must orbit the sun, 2) is massive enough to have formed a sphere by its own self-gravity, and 3) must have cleared its orbit of all other significant mass. Pluto is fine with the first two, but point 3) is where it fails. It turns out that Pluto is part of a body of Solar System objects called the Kuiper Belt, and several of these are within the orbit of Pluto and as massive as Pluto.

Within the past two decades, several new "planets" have been found within the orbit of ~30 AU to 100 AU. The largest of these is Eris, which is larger than Pluto, and was the cause of the chaos over planet definitions. Most of these objects are made of ice, which includes solid CO_2, N_2, and other things that are cold. Many have moons, and many also have large inclinations with respect to the plane of the solar system. Most orbits are highly elliptical as well.

The Kuiper Belt is a region originally hypothesized by Gerard Kuiper to exist beyond the orbit of Neptune that was the source of the *short-period* comets, or those comets that have orbits between Jupiter and Neptune. The existence of the Kuiper Belt has since been verified.

In contrast, the Oort cloud (named after Jan Oort) has not been verified. The Oort cloud is the hypothetical mass of material that is the source region for the long-period comets. The Oort cloud extends from about 100 AU to 50000 AU, which is the extent of the gravitationally bound material in the Solar System.

Chapter 22 Practice Problems

1. Oxidants are found on Europa's surface, and include O_2 and H_2O_2. These oxidants may be covered by water, and slowly released to the underground ocean. When they meet the ocean, they can react with reducing agents such as H_2S. The reaction may be:

$$H_2S + 2\,O_2 \rightarrow H_2SO_4$$

What would happen to the pH of the water if this happens?

2. How might you determine whether the water jetting out of Enceladus was in contact with rock or not?

3. Solid nitrogen ice is found on Triton's, and is the main form of elemental nitrogen. Additionally, nitrogen air makes up its atmosphere. If the ΔG of N_2(gas) is +0.861 kcal/mol at 40 K, what pressure of gas (in Pascals) would you expect as the atmosphere of Triton?

Chapter 23. Meteorites

23.1 What is a Meteorite?

A meteorite is a rock from space that has traveled through the earth's atmosphere. Most meteorites originate from asteroids, the small chunks of rock that are smaller than planets that are found throughout the solar system. A meteorite differs from a meteor and meteoroid based on its position relative to the earth and its atmosphere (Figure 23.1).

Figure 23.1 Meteor/oid//ite names depend on their location relative to the earth. A meteoroid has yet to enter the earth's atmosphere, whereas a meteor is blazing through it. If the rock survives this process and reaches the surface of the earth, it can be recovered as a meteorite.

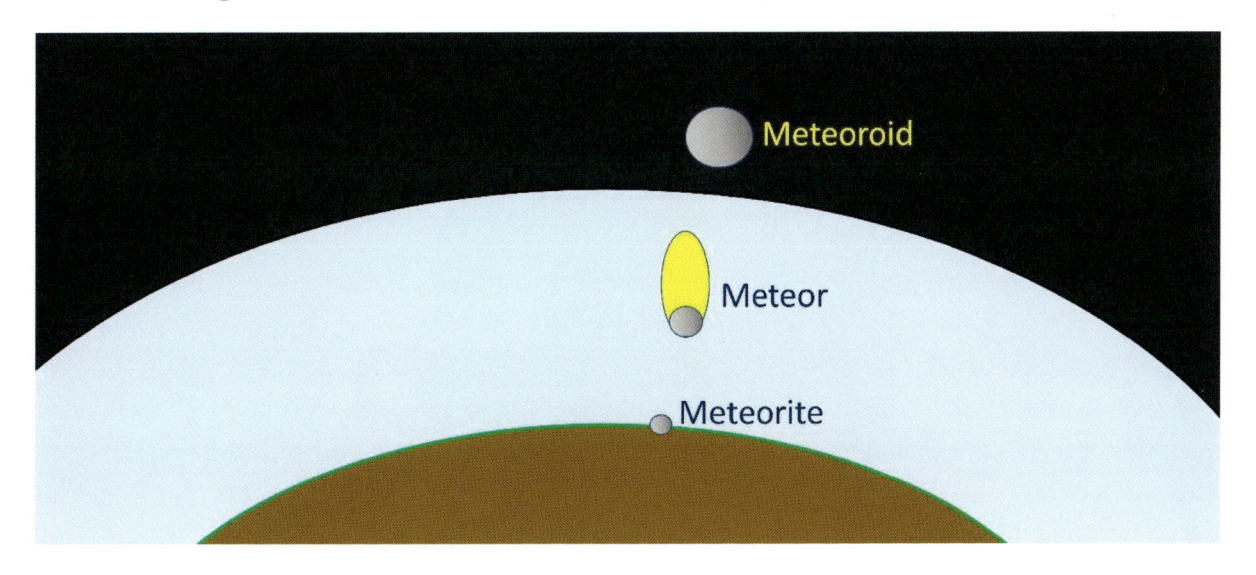

23.2 Types of Meteorites

Meteorites are grouped in different ways, depending on what is being described. All meteorites are either *falls* or *finds*. A *fall* is a meteorite that was observed to fall by human witnesses and collected subsequently. There are about 1-10 falls each year across the globe, though there are thousands of meteors. Most just aren't recovered. A *find* is a meteorite not observed to fall but collected as an "odd rock" by amateurs or by meteorite experts. Meteorites that come from Antarctica or the Saharan desert are almost exclusively find meteorites. There are about 30,000 known meteorites, and possibly many more that remain unclassified coming from deserts.

Meteorites are also separated based on composition. *Iron*, *Stony-Iron,* and *Stone* meteorites are the largest of the compositional classifications. Iron meteorites are the ones that are often most familiar to novices, as they are large chunks of iron-nickel metal that are quite different than most earth rocks. Stony-irons are about a 50:50 mixture of silicate/oxide rock and metal. Stony meteorites are meteorites that are mostly silicate in composition, but often have some metals as well. About 95% of all meteorites that fall to the earth are stony meteorites. In contrast about half of the mass of meteorites on the earth today are iron meteorites. This disparity is because massive iron meteorites make up the biggest meteorites as they stick together well.

Iron meteorites are mostly composed of iron and nickel metal. They are actually the most compositionally diverse of the meteorites with over 100 separate meteoroids sampled as part of the iron meteorite collection. However, in most cases they are made of >90% iron and nickel, often with some sulfur, phosphorus, carbon, and silicates. Most iron meteorites are probably samples of asteroid cores.

Stony-iron meteorites are the least common of the 3 major classes, and have less diversity. The two major subclasses of stony-iron meteorites are the pallasites and the mesosiderites. Pallasites are rare mixtures of metal and silicate (typically olivine) that likely originate from the core-mantle boundary of asteroids. Mesosiderites are finer-grained mixtures of these two and may be impact breccias.

Stony meteorites provide some of the best information on the origins of our solar system and on outer space as a whole. They are broadly classified into classes: chondrites and achondrites. Chondrite meteorites bear chondrules, small (~1 mm) silicate spheres that are among the most primitive matter in the solar system. Achondrites do not have chondrules. The chondrite meteorites include the Ordinary chondrites, which are the most common meteorite to fall to the earth (85% of all falls). They also include carbonaceous chondrites that may bear organic material from space, and these meteorites also define the composition of the solar system (Figure 3.2 comes from analyses of these meteorites). Achondrites are primarily crust and mantle rocks from asteroids, but include meteorites from the Moon and Mars.

23.3 How a Meteorite Falls

When a meteoroid get caught in the gravity of the earth, it accelerates to the earth's surface, and when it enters the atmosphere it becomes a meteor. Large meteors lose energy through *ablation*, the process by which mass is lost during atmospheric entry, resulting in energy loss. Most meteorites began as meteoroids that were 10-100 times as massive as when they are recovered.

A meteoroid enters the earth's atmosphere at a speed at least equivalent to the escape velocity of the earth:

$$v_e = \sqrt{\frac{2GM}{r}}$$

where v_e is the escape velocity, G is Newton's gravitational constant (6.67×10^{-11} in SI units), M is the mass of the planetary body (5.97×10^{24} kg for the earth), and r is the distance away from the center of mass of the body (effectively the radius of the earth, or 6370000 m). For the earth, this ends up at about 11.2 km/s.

An object going 11.2 km/s has a kinetic energy per unit mass equal to 6.3×10^6 J/kg, which is a lot. Objects will explode with less energy. Thus, a meteor must slow down before it reaches the earth. It does so as a result of air resistance. As a meteor travels through the atmosphere, it ablates mass, which removes energy by vaporizing the rock and slowing down the meteor (Figure 23.2). The average meteor loses 90% of its mass, causing it to slow from 11200 m/s to about 30 m/s.

Figure 23.2. Changes in mass and velocity for a 100 kg object entering the atmosphere at escape velocity.

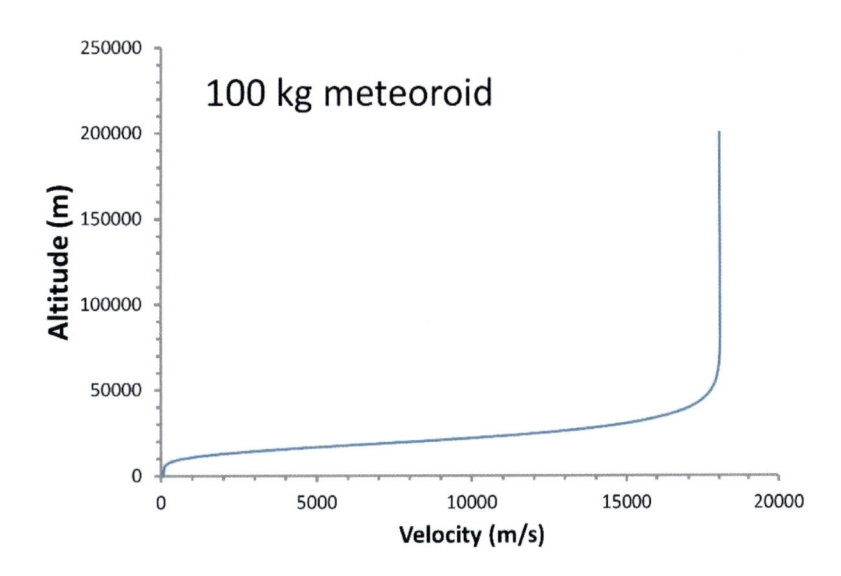

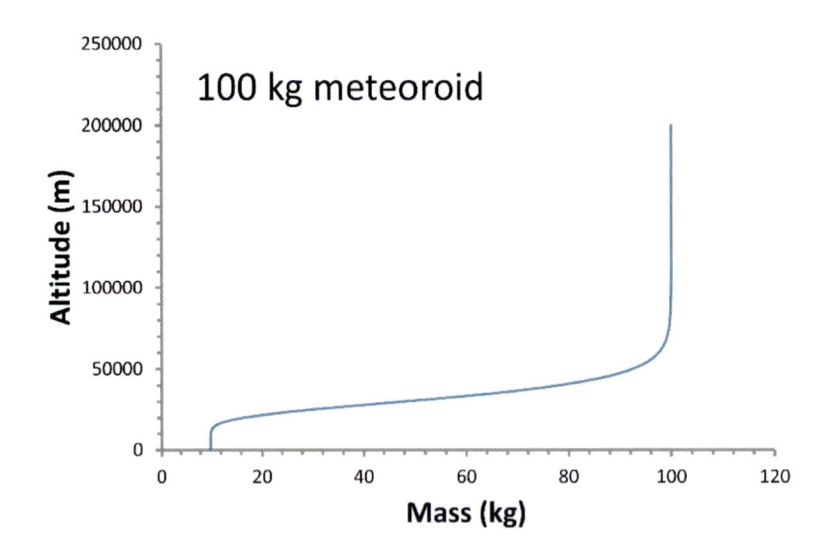

23.4 Myths of Meteorites

Because they are strange, come from space, and are mysterious, there are several myths about meteorites. Most of these can be eliminated with a bit of logic, but if you ever deal with the public, several will be common.

Myth 1. <u>Seeing a falling star means it probably struck a neighbor's yard.</u> Most meteor ablation occurs at a height of about 10-50 km (see Figure 2), and hence this is where a meteor will shine most. If you see a meteor streaking at that point, you are seeing an object ~25 km in the sky burning up. The trigonometry of this event usually implies that a meteorite is not going to be found near to the place you saw a meteor. Additionally, most meteors are as small as dust grains, which means few will be found.

Myth 2. <u>Meteorites are associated with craters.</u> Indeed, some craters have meteorites associated with them (or at least the youngest ones do), but most craters have no associated meteorites, and most meteorites are not found in craters. In order to form a crater, a meteor must not be fully decelerated by ablation so that it still has a significant amount of kinetic energy when it impacts the earth. If a meteor is slowed to 30 m/s, that is much too slow to form a crater. Similarly, most meteors are going slow enough not to cause a sonic boom if they travel overhead.

Myth 3. <u>Meteors are hot when they land.</u> Most meteors lose their energy through ablation, not heating. As a meteor loses its energy, its internal temperature remains roughly the same as it was before it entered the atmosphere. Also since most meteorites are small, they equilibrate rapidly with the surroundings, and most will be the same temperature as the environment.

Myth 4. Meteorites are radioactive. It is certainly true that space is radioactive. We benefit from a thick atmosphere that absorbs most of the radiation from space. The surface of most meteoroids is quite radioactive as they do not have atmospheres or other shielding material. However, consider again the process of ablation. 90% of the mass is lost during atmospheric entry, and as a result, there is little chance for the radioactive crust to survive entry.

Myth 5. Meteorites are made of weird stuff. For the most part, meteorites are made of the same thing as the earth (silicates) with some iron metal. Iron metal is unusual due to the oxidation state of the earth's surface. However, there are no weird elements in meteorites as the chemistry in meteorites is the same as chemistry on the earth. Nonetheless, there are several minerals that are found only in meteorites and are usually not reported elsewhere on the earth.

23.5 Meteorite Identification

The accurate identification of meteorites is a useful skill for geologists, as the public will often bring in rocks that are believed to be meteorites. Only a thorough scientific analysis can conclusively identify a rock as a meteorite. If your rock has a high probability of being a meteorite, analysis by a university laboratory is likely the next step in identification. This section of the text will inform you of the proper technique for identifying a meteorite. Not all university labs are familiar with meteorites, though geology departments will be your best bet for the proper identification of a meteorite.

In order to be officially recognized as a meteorite, either 20% of the mass or 20g of mass of the meteorite, whichever is less, must be placed via donation into a meteorite repository. This allows the mass of the meteorite to be investigated for scientific purposes if the need should arise. Additionally, the meteorite must receive a thorough mineralogic, petrologic, and compositional analysis to be classified as a specific type of meteorite.

In order to conclusively show a rock is a meteorite, and especially if the rock is stony, the rock will need to be cut to view the fresh interior. Sometimes the rock will also have to have a thin section made from it, which is a thin cut glued to a glass slide that allows for analysis of the rock by microscope. If the rock is metallic, most of the time the metal will need to be analyzed chemically to determine whether or not the rock is a meteorite, and also to provide the classification of the meteorite.

The following list provides information for identifying a rock as a meteorite. The first three points will identify about 99% or more of all meteorites. Less than 1% of all meteorites are difficult to identify as they may lack chondrules, metal, and a fusion crust, and hence these must be identified by a series of expensive chemical and isotopic tests. If the rock brought to you falls in this last 1% of rocks, it will often not be analyzed because of the extremely high likelihood of being a terrestrial rock. The rock will typically be analyzed only if there is justifiable reason for analysis, for instance coming from a location with known meteorite falls, such as northwest Africa. If you are provided information on the location where you found the rock, this may also help determine whether these tests would be useful.

Fusion Crust: The most obvious indicator of a meteoritic origin for a rock is the presence of a fusion crust. A fusion crust forms as the meteor flies through the atmosphere, where it burns up by the process of ablation described above. The fusion crust will usually be black, although it weathers to brown or rusty red with time. Larger meteorites, with masses greater than 1 kg, may also have regmaglypts, which are finger-sized depressions resulting from chipping of the rock during entry (Figure 23.3). Not all meteorites will have a fusion crust, if a meteorite explodes during entry close to the ground, the exploded bits may reach the surface without substantial formation of fusion crust. The fusion crust of other meteorites may weather or break off with time, hence these meteorites must be identified using other methods. The fusion crust of iron meteorites is less obvious than of stony meteorites, and these meteorites also weather quickly, hence only about 90% of all meteorites show some fusion crust.

Figure 23.3. This ordinary chondrite is covered in regmaglypts.

Chondrules: About 90% of all meteorites that fall are chondrites, most of which have chondrules. Chondrules may be visible on the exterior of a meteorite, and range in size from about 0.1 mm to 2 mm in diameter. Most chondrules should be visible on a fresh cut surface or in thin section, if they are present. These chondrules are typically composed of mafic minerals such as olivine, pyroxene, and feldspar (or feldspar glass). Other minerals may include sulfides, such as troilite, iron metal, and metal oxides such as chromite, among other rarer minerals (including SiO_2 and phosphates).

Mineralogy: Meteorites are characterized by minerals that are rare to virtually non-existent on the earth. The presence of iron metal is an excellent identifier of a meteoritic origin of a rock. About 90-95% of all meteorite falls will have significant iron metal. This iron metal is usually present as the mineral "kamacite", which is a crystalline form of iron. Kamacite is strongly attracted to a magnet and should also be easily visible in a cut section of a meteorite, or sometimes may be visible on the surface. If there is enough of this mineral, then etching this phase may cause the Widmanstatten pattern to become visible, which is caused by the differences in etching of kamacite and the more nickel-rich taenite. Other key mineralogic indicators of a meteoritic origin for a rock include the sulfide troilite (FeS), the phosphide schreibersite [$(Fe,Ni)_3P$], glass with a feldspar composition ($NaAlSi_3O_8$, $KAlSi_3O_8$, $CaAl_2Si_2O_8$), and others which are often type-dependent. Constraining the mineralogy of a meteorite is an important step in its classification.

Ni abundance. Most meteorites are highly enriched in the element nickel (element symbol Ni) relative to terrestrial rocks. The nickel abundance is approximately 5% or greater of the iron metal abundance in a given meteorite. Note that over-the-counter nickel tests, such as those designed to identify metals capable of irritating a nickel allergy, will not provide definitive results. Other chemical assays may be required for determination of reasonable amounts of nickel.

Composition: Frequently, an elemental analysis of meteorite can be used to both differentiate a meteorite from a terrestrial rock, and also classify the meteorite according to type. Meteorites will tend to be enriched in the platinum group elements (Ru, Rh, Pd, Os, Ir, Pt at ppm concentrations) relative to terrestrial rocks, and will be comparably depleted in many lithophile elements, such as the rare earths. The abundance of iridium, nickel, and germanium are critical to identifying type for iron meteorites, and the composition of various minerals (for instance, manganese content of olivine) can help elucidate the origin of a stony meteorite.

Oxygen isotope composition: One of the strongest identifiers of a meteoritic origin of a rock, and also one of the least easy to perform, is an oxygen isotope analysis. Most meteorites have oxygen isotope ratios, which are the ratios of $^{17}O/^{16}O$ and $^{18}O/^{16}O$, which differ from the terrestrial fractionation line (see Figure 9.1). The amount of difference is diagnostic of the type of meteorite. Unfortunately, few labs can do three oxygen-isotope analyses.

— So Uncle Earl thought he found a meteorite

— Is that so?

— Yeah, so he took it in to the local University

— Is he rich now?

— Nope - it turns out it was a Meteor wrong!

Chapter 23 Practice Problems.

1. What is the escape velocity of Mars?

2. Are these rocks meteorites:

Chapter 24. Instrumentation

24.1 Instrumentation as a Science

As we have said several times before, we will cover this topic only in a cursory manner. The science behind analytical chemistry can build an entire career. The detection and identification of different elements, their isotopic composition, and their speciation are all critical to chemical analyses of natural samples, and as new tools arise to detect and ID materials, having these tools available for scientific and regulatory analyses strengthens our understanding of how things work.

However, no matter where you are, not all tools will be available for you to use, especially in the field. Some are too complex, some are too expensive, and some are too specialized to merit purchase. As a result, knowing when to use specific tools at certain points in your career will be of great benefit.

24.2 Considerations for instruments

Before setting out to solve the great geochemical mysteries using analytical chemistry tools, you need to figure out a few things. First, and most important is "What do you want to do?" Do you need to measure speciation or abundances? Are you looking to analyze a pure sample or are you hunting for very dilute material in a substantial matrix? Are you looking at a solid, a liquid, or a gas? Then, consider, "How sensitive do you need your analysis?" Are you looking for percentage determinations? Part per million? Part per trillion? Just a simple mineral ID? Third, what sample preparation do you need to do? Do you need to make a thin section? Powder the sample? Fourth, how much do you have to do it? There are two significant costs with most analyses: how much the sample prep costs (mounting a rock cost money, as does dissolving one), and how much the instrument costs. If you're not buying the instrument because you're not using it regularly, then how much does an individual sample cost to analyze? Finally, in some cases having a rugged instrument means you can take it to the field. How important is portability?

24.3 Types of Analyses

An overview of several analytical tools used by aqueous geochemists follows. This is by no means an exhaustive list, but provides a basic overview of the tools available for a field geochemist or environmental chemist. These range in price, portability, sensitivity, and precision by a significant amount. Table 24.1 outlines many of these features. Many of these techniques use light in some form for analysis. Light is useful as many compounds absorb light at specific wavelengths (very similar to why different objects have different colors). Additionally, it is easy to separate light using a prism or similar tools.

Electrodes. One technique that is used frequently in water analysis involves electrodes. Electrodes measure voltage potentials between a fluid and a standard, providing new data. pH meters are a variety of electrode, as are redox potential meters. Both compare the voltage potential of a sample with a standard, often a "standard calomel electrode" or a "standard hydrogen electrode". Similar to these is a TDS meter that measures the resistivity or conductivity of fluid, which is typically directly related to the amount of solutes dissolved within. Electrodes used in analysis are typically small, portable, and cheap.

Optical Spectrophotometry. A common tool in field geology and hydrology are portable spectrophotometers. These devices measure the absorbance of light by a sample of water that in turn provides a constraint on concentration. To do so, the water sample is typically filtered, then a reagent is added to the sample. The reagent reacts with the species of interest (for instance, phosphate), forming a chemical complex with a specific color. If more of the compound is present in the water, then more of the reagent will react and will enhance the color of the solution. This process follows a law called "Beers Law". Beers Law states:

$$A = \varepsilon \lambda c$$

where A is the absorbance of the solution, ε is a unique absorptivity of the reagent and compound of interest, λ is the width of the sample chamber, and c is the concentration of the compound of interest. This equation states that the amount of light absorbed at a specific wavelength (tuned to match a specific absorption feature of the reagent and compound of interest) is a function of the concentration of the compound of interest, a specific absorbance constant of the compound and reagent, and the width of the sample cell. This method can be quite accurate over small range of sample concentration, provided that the concentration of the compound of interest does not vary by about a factor of 10 to 100 between samples. Sometimes other ions may compete with the compound of interest (for instance, arsenate for phosphate), and cloudy water will almost always give bad results, but if these are accounted for, then the analysis can be accurate, cheap, and portable. This method is generally used to identify solute concentrations and occasionally speciation in a solution.

ICP-OES. This technique stands for Inductively Coupled Plasma Optical Emission Spectroscopy. ICP is a method that takes a sample and heats it until it becomes plasma, or ionized gas. At this temperature molecules cannot easily form, and hence everything is effectively atomized. These atoms are heated to the point where they become gaseous ions. Ions gain and lose electrons according to set energy levels (e.g., the Bohr model of the atom), and hence each element can be analyzed for according to its unique set of energy levels, corresponding to unique light wavelengths. ICP-OES can be used to quantify major elements, for instance in a rock sample.

This can allow you to determine what a rock might be based on its composition. ICP-OES is not portable, and it is expensive. Sample prep time can be significant and complex.

ICP-MS. This technique stands for Inductively Coupled Plasma Mass Spectrometry. Mass spectrometry is a tool that separates materials according to their masses and their charge. This separation is done by a variety of processes including based on time of flight (heavier things take longer than lighter things to travel an equivalent difference) as well as mass to charge ratio. This method can be extraordinarily sensitive, getting to part per trillion level detection limits when well calibrated and tuned. It is primarily an elemental analysis method. Any sort of solution can be analyzed by this method, although if you intend to determine the trace element (for instance, rare earth elements) composition of a rock, you must first dissolve it. Alternatively, a laser can be placed before the ICP-MS, and can be used to ablate material off of a rock slide, providing a point-by-point analysis. An ICP-MS is capable of analyzing the composition of a solution (including a dissolved rock) to extremely low levels, but is expensive, not portable, and can require training to use properly.

GC-MS. A Gas Chromatography Mass Spectrometer uses a mass spectrometer in a fashion similar to the ICP-MS, but instead of analyzing for elements it analyzes for molecules. The gas chromatography column separates gas molecules based on their chemistry, and there is a whole science dedicated to the separation of things, called chromatography. A GC-MS is especially useful for organic compound analysis, especially when the organic compounds are mixed and have abundances less than percentages. Recently GC-MSs have become small and portable enough to take in the field, in addition to becoming much less expensive. However, a laboratory with a dedicated GC-MS will have much greater limits of detection and utility than a portable GC-MS.

NMR. A Nuclear Magnetic Resonance spectrometer is a tool that is primarily used in chemistry and pharmaceuticals. It mostly works for NMR-sensitive elements, which include H, C, F, and P, with a few others. Geologists or environmental chemists who are studying these elements may employ an NMR at some point in their careers. NMR provides detailed chemical data on the chemistry of these elements, but these elements should be present at millimolar concentrations or higher, and mixtures can present problems. The method is not portable, and is quite expensive. Spectral interpretation also takes some training.

XRD. X-Ray Diffractometers have been used for close to one hundred years in geology, and are a standard tool in mineralogical analysis. XRDs use X-rays to measure the crystal lattice spacing of a crystalline solid, and these spacings are typically unique to a specific mineral. These measurements follow Bragg's law:

$n\lambda = 2d \sin \theta$

where n is an integer, usually 1 or 2, λ is the wavelength of x-ray being used (usually known well before-hand and specific to the XRD being employed), d is the spacing and is usually the data product desired, and θ is the angle of incidence for the X-rays, and is measured accurately during analysis. Most of the time XRDs require powdered mineral samples, and pure sample powders are typically easier to analyze than mixtures. With mixtures, compounds can be identified at about the 5% level. XRDs have become cheaper recently, but aren't really portable, though they are easy to use once trained.

EPMA. The Electron Probe MicroAnalyzer is an extremely sensitive tool. It is effectively a scanning-electron microscope with a well-calibrated elemental analyzer attached. It can analyze single points on a slide that are as small as about 1 μm^2. From this point it can give an elemental analysis down to the tenths of a percent level accuracy, sometimes better. This can be useful to identify specific minerals within a rock sample, as well as identifying melts, glasses, heterogeneous materials, and many other things. If an analysis of a solid would benefit from spatially-resolved chemical analysis, this is the ideal tool. It is extremely expensive, not in the least portable, and requires a trained technician to use and maintain it.

Chapter 24 Practice Problems

1. Let's say a mining company hires you to investigate core samples to determine where nickel ore might be. In order to be an ore, the rock should have at least a part per thousand nickel, and you have several thousand meters of core samples. How would you do this analysis?

2. The company now wants you to identify the primary nickel-bearing mineral within this ore sample. Assume that the ore is not all nickel, but include non-nickel bearing minerals. What tool would you use?

3. After mining all the juicy nickel out of the ground, the government comes in and wants to make sure the mining company hasn't polluted with groundwater with organic contaminants during the mining operation. What tool might the government use to investigate the extent of pollution?

Method	Application	Utility	Portability
Electrodes	Quantifying abundances of specific elements	Specific	Great
Optical Spectroscopy	Quantifying abundances of specific elements	Moderate, but abundant	Good
NMR	Identifying compounds, some quantification	Great for H, C, F, P	Poor
ICP-OES	Quantifying major element abundances, some minors	Good, though not sensitive for low abundances	Poor
ICP-MS	Quantifying trace element abundances, some majors	Good, though not for majors	Poor
XRD	Identifying minerals, other crystalline compounds	Only most abundant compounds	Poor-moderate
EPMA	Identifying minerals/compounds in thin section	About 0.01% abundance and greater	Poor
GC-MS	Identifying compounds, some quantification	Great for Organics	Moderate

Method	Cost (equipment as a whole)	Cost (per Sample)
Electrodes	Cheap (100-1K)	Cheap
Optical Spectroscopy	Cheap (1K)	Cheap
NMR	Expensive (100-4000K)	Cheap
ICP-OES	Expensive (200-500K)	Moderate (sample prep)
ICP-MS	Expensive (200-500K)	Moderate (sample prep)
XRD	Cheap-Expensive (5K-500K)	Cheap
EPMA	Expensive (1M+)	Moderate (sample prep), analytical time
GC-MS	Cheap-Expensive (50K-1M)	Cheap

Printed in Great Britain
by Amazon